AF609498

Swoop Sing Perch Paddle

BIRDS BY
Carry Akroyd

WORDS BY
John McEwen

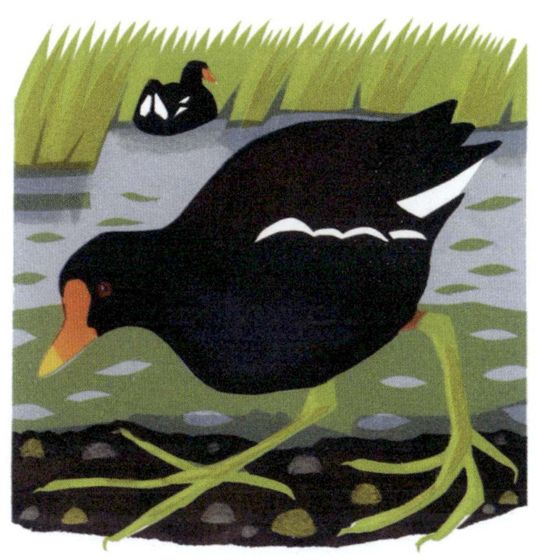

BLOOMSBURY WILDLIFE
LONDON • OXFORD • NEW YORK • NEW DELHI • SYDNEY

For Gordon
who cooks dinner and
cares for Bears

David, Jo,
Duncan, Una, Iris,
Innis, Oscar, Mabel

BLOOMSBURY WILDLIFE
Bloomsbury Publishing Plc
50 Bedford Square, London, WC1B 3DP, UK
29 Earlsfort Terrace, Dublin 2, Ireland

BLOOMSBURY, BLOOMSBURY WILDLIFE
and the Diana logo are trademarks of Bloomsbury Publishing Plc

First published in the United Kingdom 2024

A catalogue record for this book is available from the British Library

Library of Congress Cataloguing-in-Publication data has been applied for

ISBN: HB: 978-1-3994-1802-7; ePub: 978-1-3994-1800-3; ePDF: 978-1-3994-1797-6

2 4 6 8 10 9 7 5 3 1

Designed by Austin Taylor
Printed and bound in China by
RR Donnelley Asia Printing Solutions Limited Company

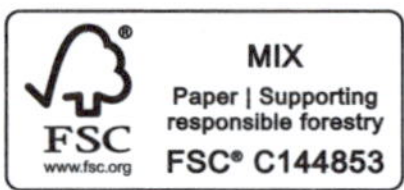

To find out more about our authors and books visit www.bloomsbury.com
and sign up for our newsletters

CONTENTS

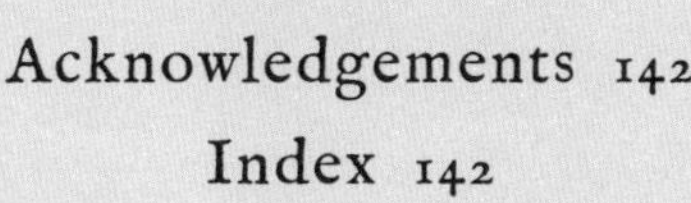

FOREWORD

Every four weeks in *The Oldie* office in Fitzrovia, I get a double burst of joy. First, John McEwen's charming, lyrical words about our Bird of the Month fly into my inbox. And then, moments later, Carry Akroyd's enchanting illustration of the same bird flutters into Oldie Towers. Every year, I ask them both, 'Oh God, are we running out of birds?' And, every year, they calm me down. No, there are plenty more. They give me the same comfort Ted Hughes supplies in *Swifts* about that heartening moment when those lovely birds return from Africa in the spring, reassuring him somehow that the globe, and creation, are still working.

Swifts fly between the covers of the book in your hands. As do so many British favourites: the tawny owl with his wide-awake eyes at midnight; the red kite, wings aloft, always apparently perched on a thermal; the miraculously bright blacks, whites and reds of the puffin and oystercatcher.

John McEwen's light touch disguises his deep, intellectual understanding of birds. He captures the look and behaviour of them so succinctly. Here he is on the cormorant: 'That birds descend from dinosaurs is easy to imagine of the cormorant (*Phalacrocorax carbo*), with its reptilian neck and head, plumage demarcated like scales, stiff-quilled tail and upright carriage.' And he is literally poetic, too: here quoting Andrew Marvell on the kingfisher – 'Charm'd with the sapphire-winged mist.'

In her pictures, Carry Akroyd instantly gets the 'jizz' of the bird – its characteristic outline and movement that birders can recognise in the swiftest of glimpses in a brambly thicket at dusk.

So often, she has illustrated her subjects in exactly the poses I'm used to seeing them: the predator-alert blue tit glued to the bird-feeder with its curled talons; the treecreeper defying gravity, perched at the absolute vertical; swallows hanging around on the telephone line, waiting for autumn to call them south.

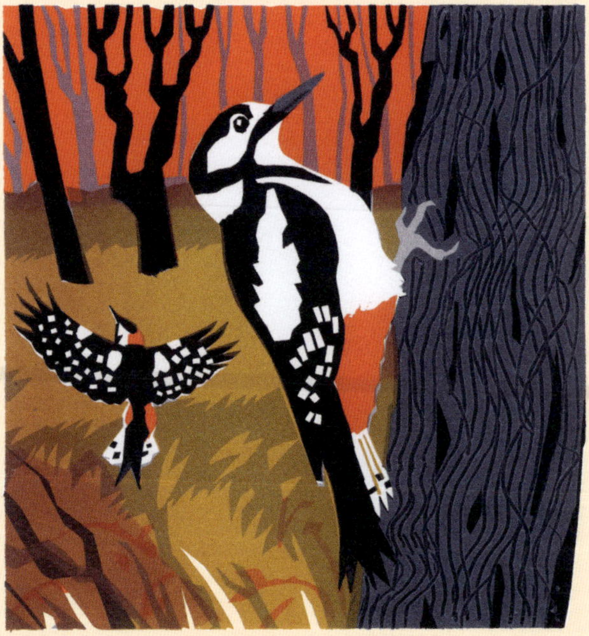

Experienced birdwatchers will feel a rush of pleasure on seeing their old friends. And, if you don't know anything about birds, this book is an instant route to knowing about them, recognising them and loving them. A passport to the kingdom of the sky.

HARRY MOUNT,
Editor of *The Oldie*, spring 2024

INTRODUCTION

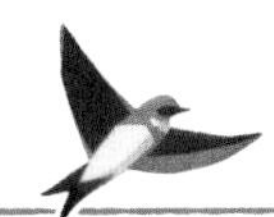

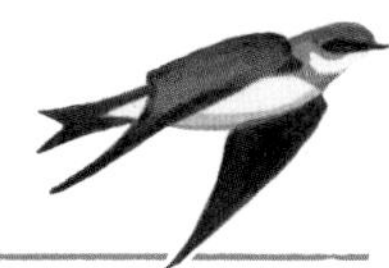

Bird books invariably use illustration as a guide, whether through paintings or photographs. Carry's screenprints are interpretations, with bold abstract solutions to convey birds swooping, singing, perching, paddling. As she is very much a landscape artist, it is always exciting to see her answer to the latest challenge of sky, land or sea, and how she captures the essence of the bird and fits it to habitat and season.

The aim of the captions is to celebrate, amuse and inform in that order. Where possible, poetry, quotation and personal insight take precedence, to circumvent general knowledge and indispensable Google. John Clare has pride of place as a bird poet by right.

In 2024 it is hard to avoid a repetitive lament when writing of birds, technology reducing country-dweller numbers as remorselessly as it has hedges, reedbeds and the rest. As for pollution, even our bodies contain microplastics. On the bright side, since 1970, of the 130 UK bird species studied in 2022, a Defra report estimated 24 per cent increased, 46 per cent showed little change, and 29 per cent decreased. Some ornithological terms require explanation: 'territories' can be held by one bird, a pair or a flock. 'Pairs' excludes failed nesters and clutch numbers. As most birds flock in winter, 'individuals' are easier to count. 'Estimate' should not be underestimated.

Age only increases the recognition of how like us in character and circumstance birds are. My mother found consolation as a widow through friendship with a barn owlet; my terminally ill wife in the welcome of a flightless Canada goose. Adrian Richmond, whose *Big Issue* pitch is the piazza of Westminster Cathedral, daily reveals the needs of the local pigeons through shepherding his feral flock, 40 of whom he knows by name. And how thrilling to receive a call from my grandson Oscar, aged 10, excitedly reporting seeing his first goldcrest. Kinship with birds reaches its apogee with Yvette Brown's charming booklet, *Bird Signs*, which reveals our personalities by the nature of the bird she allots to our birth month.

Of authors cited, Henry Douglas-Home, my Berwickshire neighbour known since childhood, was BBC radio's 'Birdman' in Scotland, hence the title of his ornithological memoir *The Birdman*. Another boyhood influence, later friend, was Denys 'BB' Watkins-Pitchford. Among authors he admired was naturalist W. H. Hudson, himself a friend of Grey of Fallodon, who wrote another bird classic, *The Charm of Birds*. The captions are stitched together by lifetime dates, a reminder of how little we have changed. Birds have been and always will be loved. For BB there was no song more beautiful than the blackbird's. For this bird lover, raised in pink-footed goose country, no chorus is more stirring than a skein of them in full cry. Thanks to their liking for discarded sugar beet tops, there are more of them than ever. I heard them with my parents and siblings as a boy. Now in Norfolk, I hear them with Frankie and Albi, my great-great-nephews:

> *Then we heard them!*
> *The tribes,*
> *the generations since the tribes,*
> *coming back.*
>
> Swarden Burn, 'Pinkfeet'

JOHN McEWEN, spring 2024

MAKING THE BIRD PRINTS

When invited to be the illustrator for John McEwen's 'Birds of the Month' column in *The Oldie* magazine, I didn't appreciate what a chunk of my life over the next 10 years I would spend thinking about making little birdie pictures. It's been fun. John plans ahead so each bird lands in the magazine at the appropriate time of year and avoids having all his ducks in a row. I endeavour to echo this variety with contrasting compositions in consecutive illustrations.

Arranged by month, this collection mixes up the diversity of bird forms and habitats rather than presenting the birds in the species order used in proper bird field guides. Bird species that are present with us all year round could have been placed anywhere in this book, and so have been arranged more randomly. Sometimes, two images just look interesting together.

As a casual birdwatcher, I first bring to mind the strongest impression in my memory of having seen that particular bird, mostly considering its context. What is its typical habitat and stance – perched or in-flight? I think of my pictures for the magazine as being ornithology for beginners, representing the bird in a simplified, recognisable form.

I prefer printmaking for illustration work, in this case, serigraphy (screen printing). Although it's a rather convoluted way of going about things, the simplification and flat colour reproduce well. I usually make between 6 and 10 small prints by hand on a tabletop in my studio, using a small mesh screen, a 7-inch squeegee and hand-cut paper stencils.

I start by making thumbnail sketches to work out the composition, then create a drawing of the right size and shape (slightly

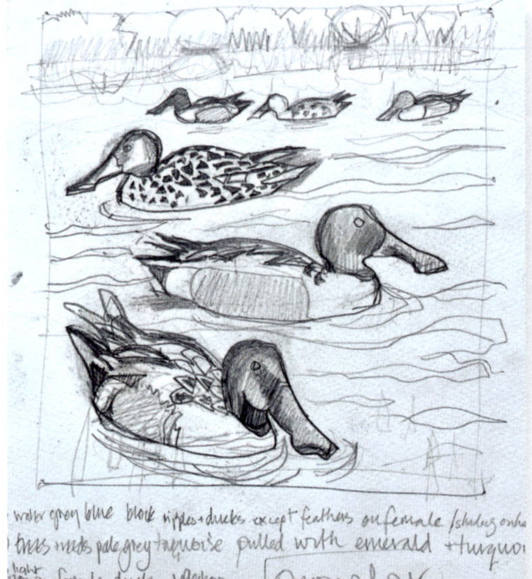

larger than the size it will be reproduced), which I use as a sort of map. At this stage, I plan how to construct the image by printing just four layers of colour. It's a puzzle to grapple with, like a jigsaw, sometimes demanding a bit of ingenuity. It's also important to consider in which order to lay the colours down.

Using tissue paper, I cut a stencil to block out areas to be left free of ink. I print this first layer of colour directly on top of my drawing map, and then with the stencil attached to the screen, I can print it again onto around ten other clean pieces of paper. I leave these to dry, clean my screen, and then start to cut out a second layer of stencil. In printing the second layer, anywhere the colour is allowed to overlap the previous colour, a third shade will be created. This doesn't always work out on the first print, and it can take two or three more colour changes to get it right.

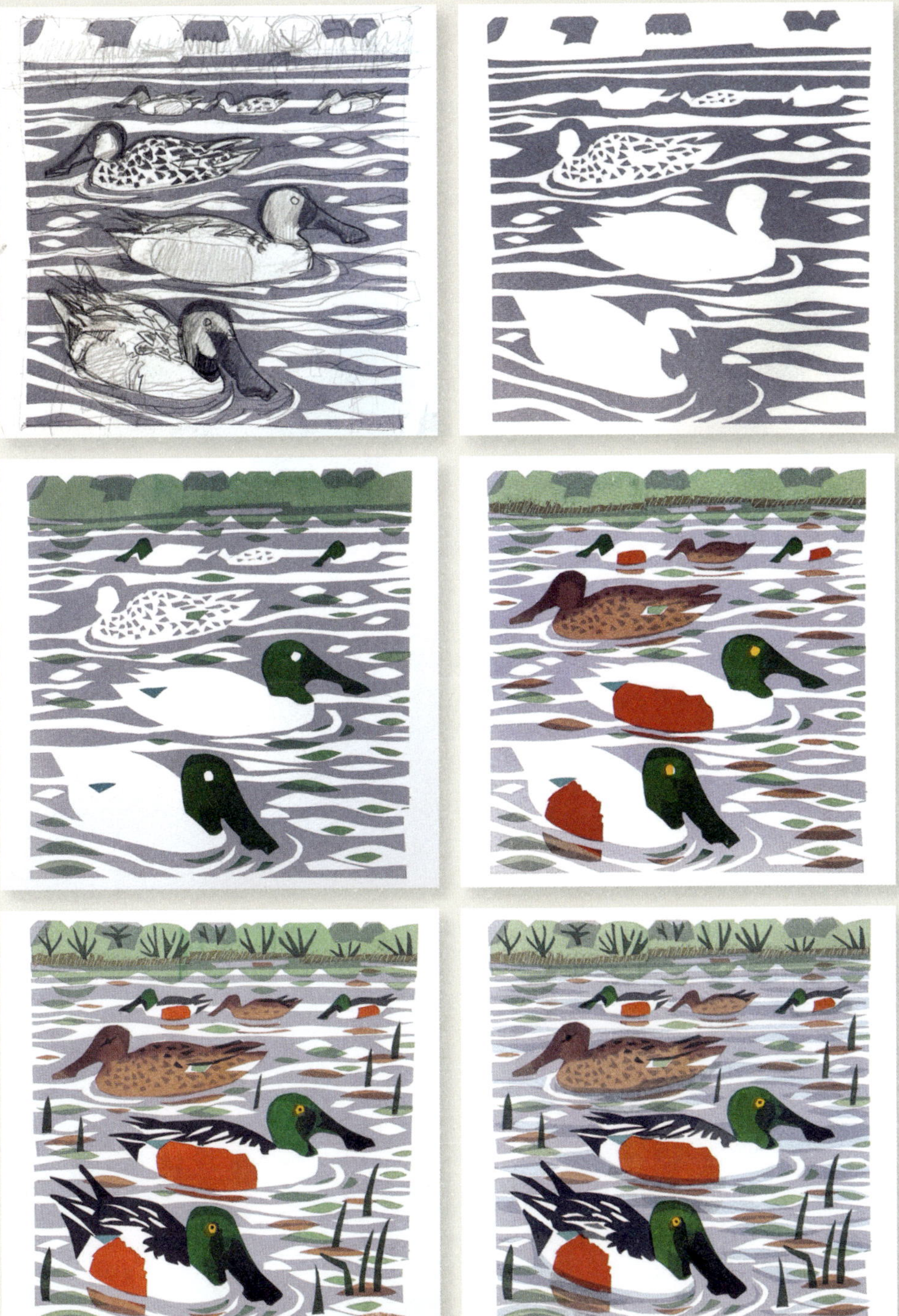

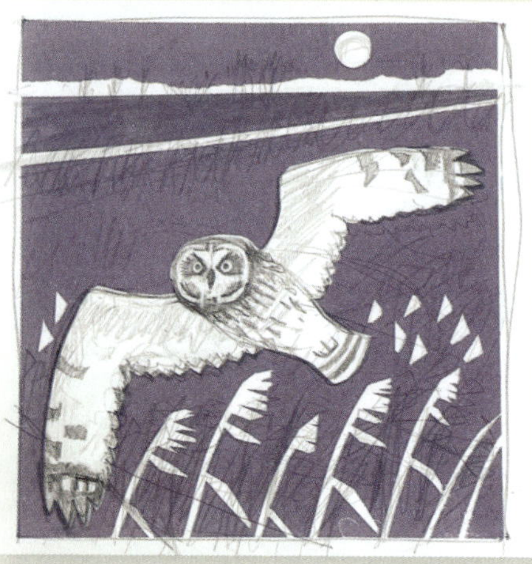

There are some extra modifications available: using a stiff brush, I can force some dark green ink into a small area of mesh before pulling a lighter green across the whole layer, creating a small variation. The blending is a little unpredictable, so no two prints in my small editions will be exactly alike. It may also take modifications over a few prints to get the colour balance right. By the time I get to the fourth layer, the image should be resolved and balanced. Occasionally, a fifth layer is required just to sort things out. One or two may end up in

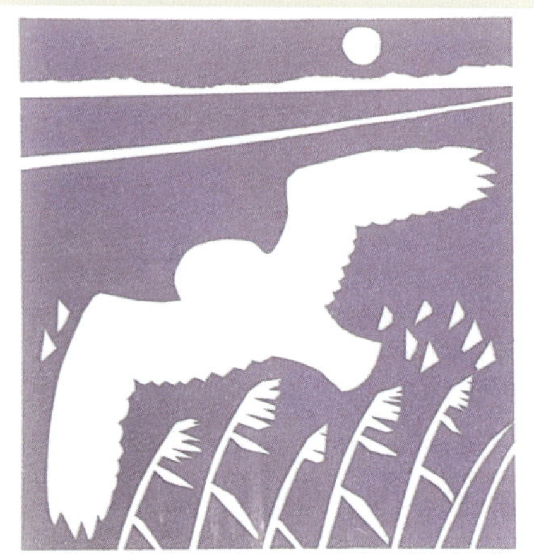

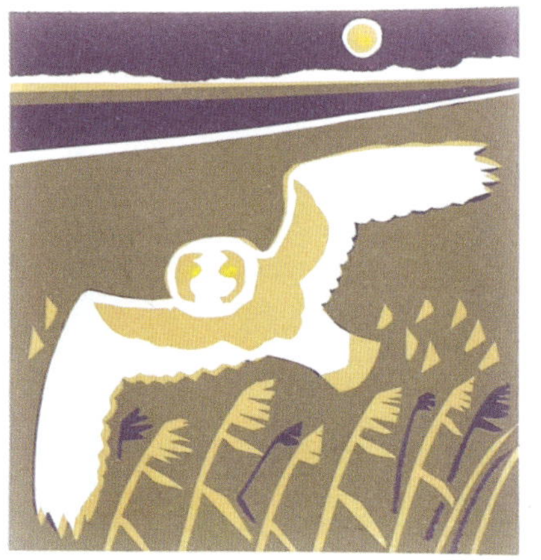

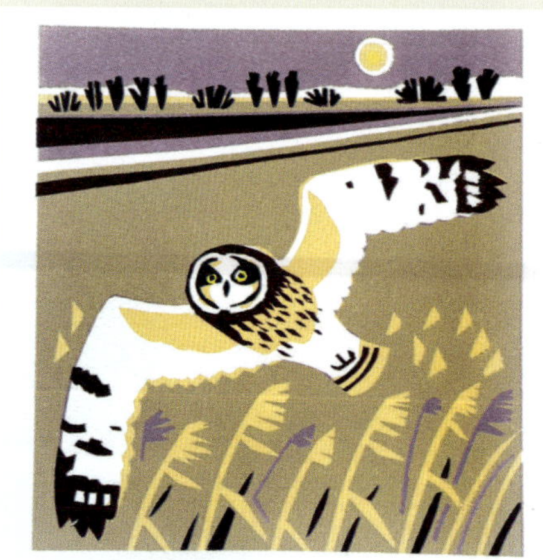

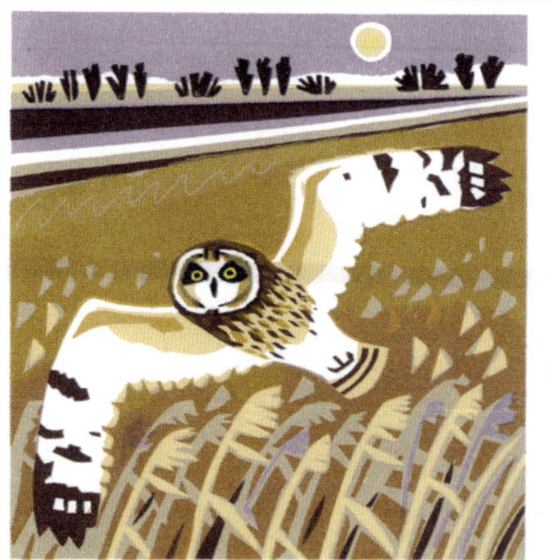

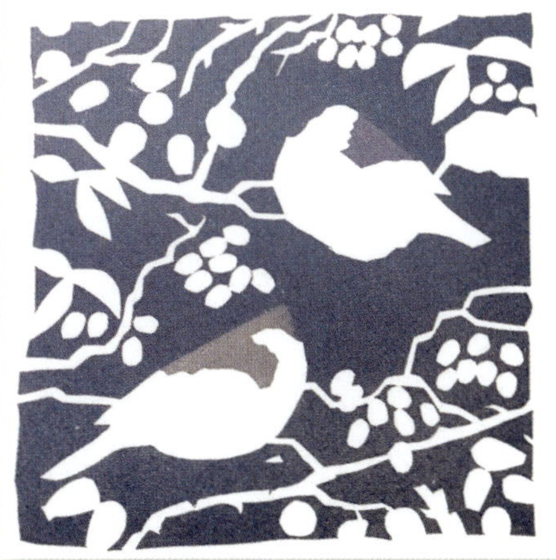

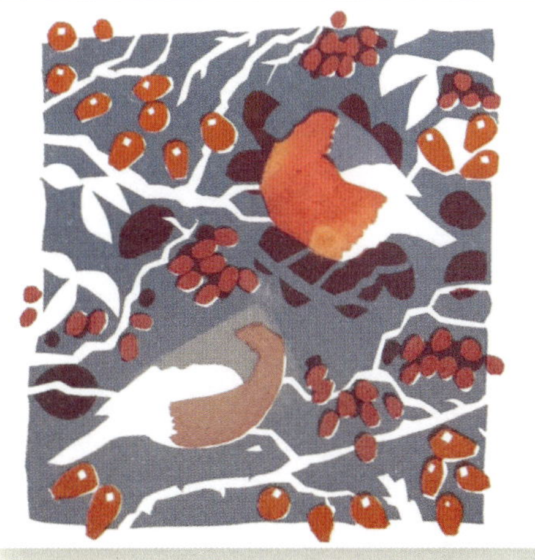

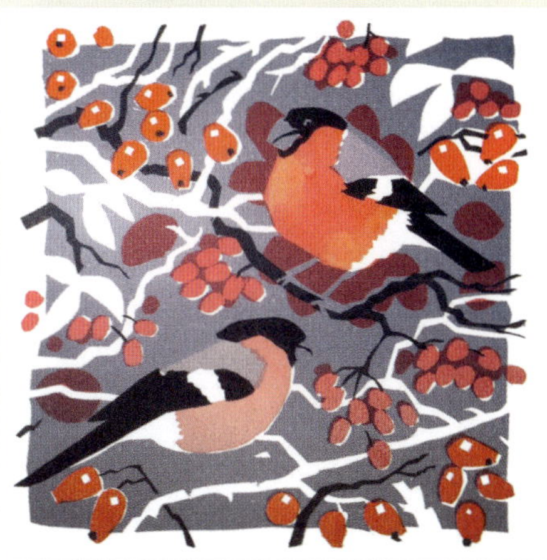

the bin. From those that don't, I pick out the best one to be digitally scanned by my technical assistant (husband), ready to send to the Art Editor at the magazine.

My binoculars have missed seeing some species. I do like to spend a little time observing to get a sense of the bird's posture and demeanour, as well as its situation. John McEwen has often brilliantly arranged for us to go on special expeditions with his friends and relations to see birds I've never seen before, such as the great bustard and the black grouse.

I have reproduced many of the little bird pictures in this book on my small-run annual calendar. Each month, John McEwen sends me his full-page column with its literary quotations, obscure facts and anecdotes. When I first planned my calendar, I persuaded John to extract a short nugget from his articles to put below each month's print, and it is these condensed snippets that he has adapted for this book.

CARRY AKROYD, spring 2024

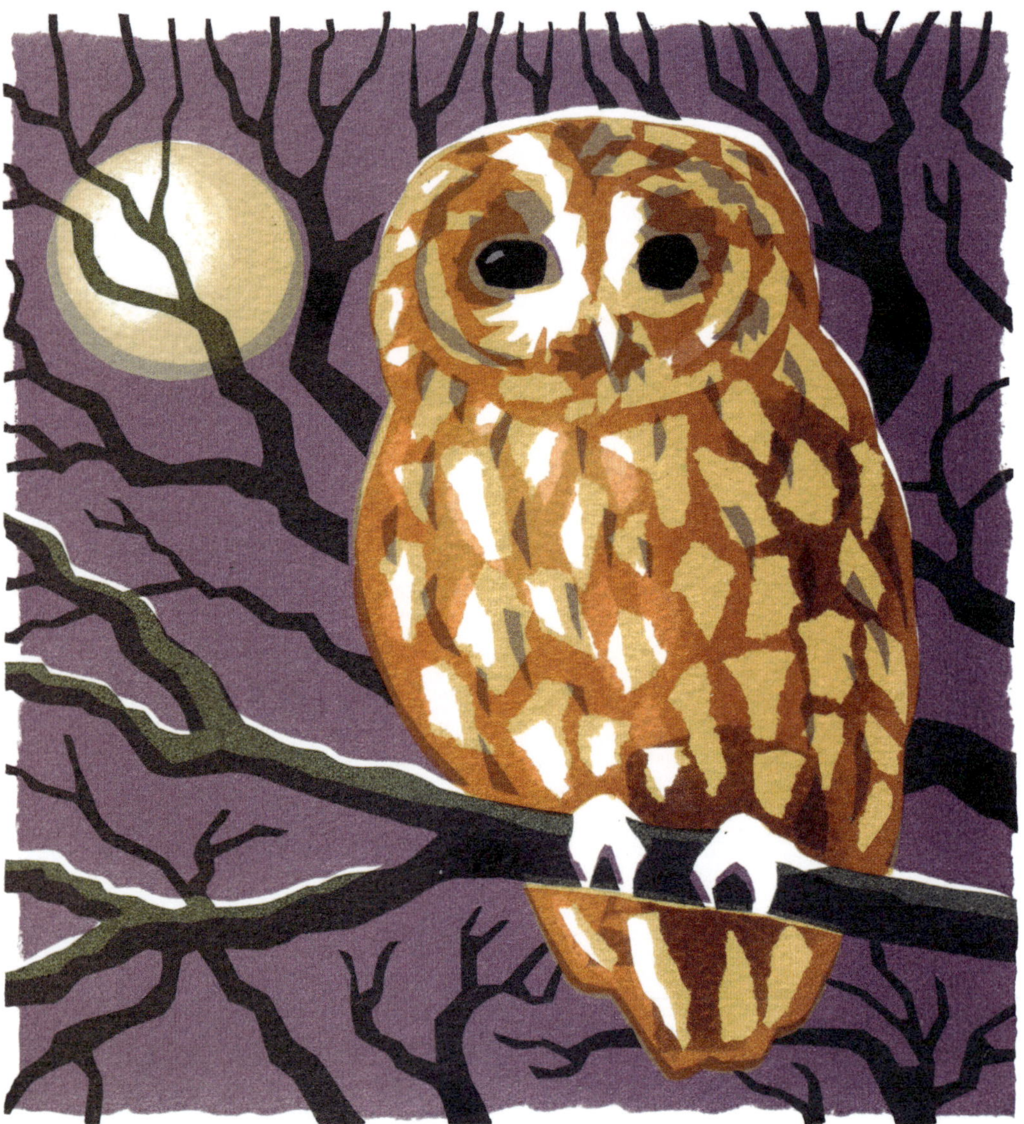

The **tawny owl** (*Strix aluco*) can be a winter nester, *tu-wit* the female's most frequent call, *tu-woo* the male's, as they mate and territorialise. Often sleepy flyers when disturbed by day, they can be lightning quick at night. Henry Douglas-Home (1911–79), a pioneer of outdoor broadcasting with BBC Scotland, often used Dundock Wood at the Hirsel, his family's ancestral home in Berwickshire, as a recording location. His brother William, the playwright, an indefatigable practical joker, was present when there was a pressing demand for a tawny owl's hoot. Several microphones were set up. To the delight of the crew, a tawny produced a brilliant call – so good BBC radio dramas often used it to convey a nocturnal scene. For many years, when Juliets leaned from their balconies or Frankensteins worked in their candlelit laboratories, it was unwittingly to the sound of William's perfect hoot.

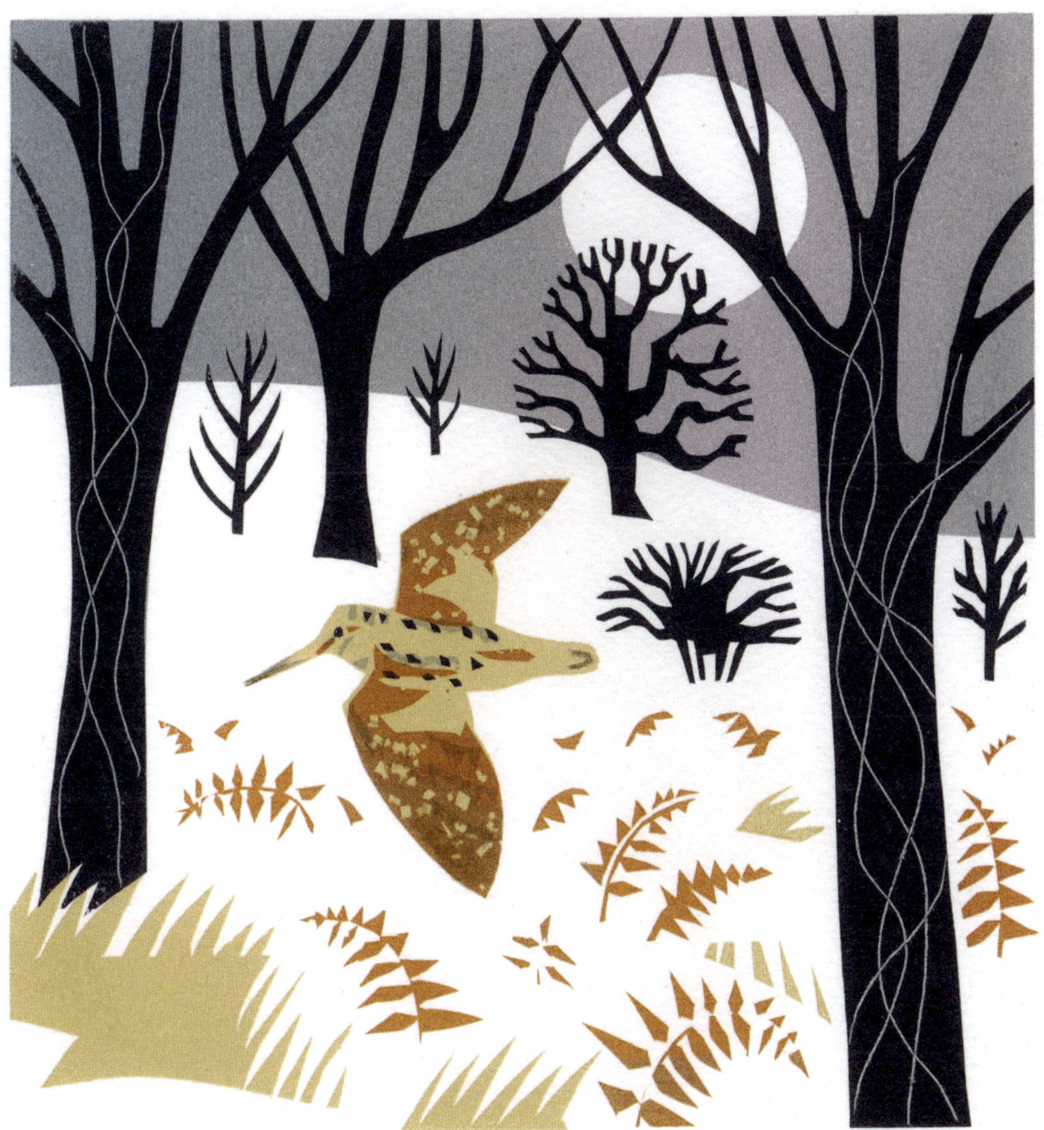

The **woodcock** (*Scolopax rusticola*) is an inland wader, hiding by day, feeding at night. White tips to its tail feathers are for display and nocturnal communication. Research initiated by Jamie Dunning, a PhD candidate at Imperial College London, proved the tips are 31 per cent brighter than the whitest feather of any other bird. Most UK woodcock are winter migrants (57,000 resident males, 1.4m migrants); they arrive in autumn in 'falls' or flocks, and tracking has revealed a migrant invariably returns to the same location. During spring dusks, males fly a 'roding' circuit of a woodland territory, repeating a satisfied grunt. Adult woodcock carry chicks to safety by flying with them clutched between their thighs. In 2022, tweets featuring 'woodcock' had a Norfolk bird charity locked out of Twitter for a week.

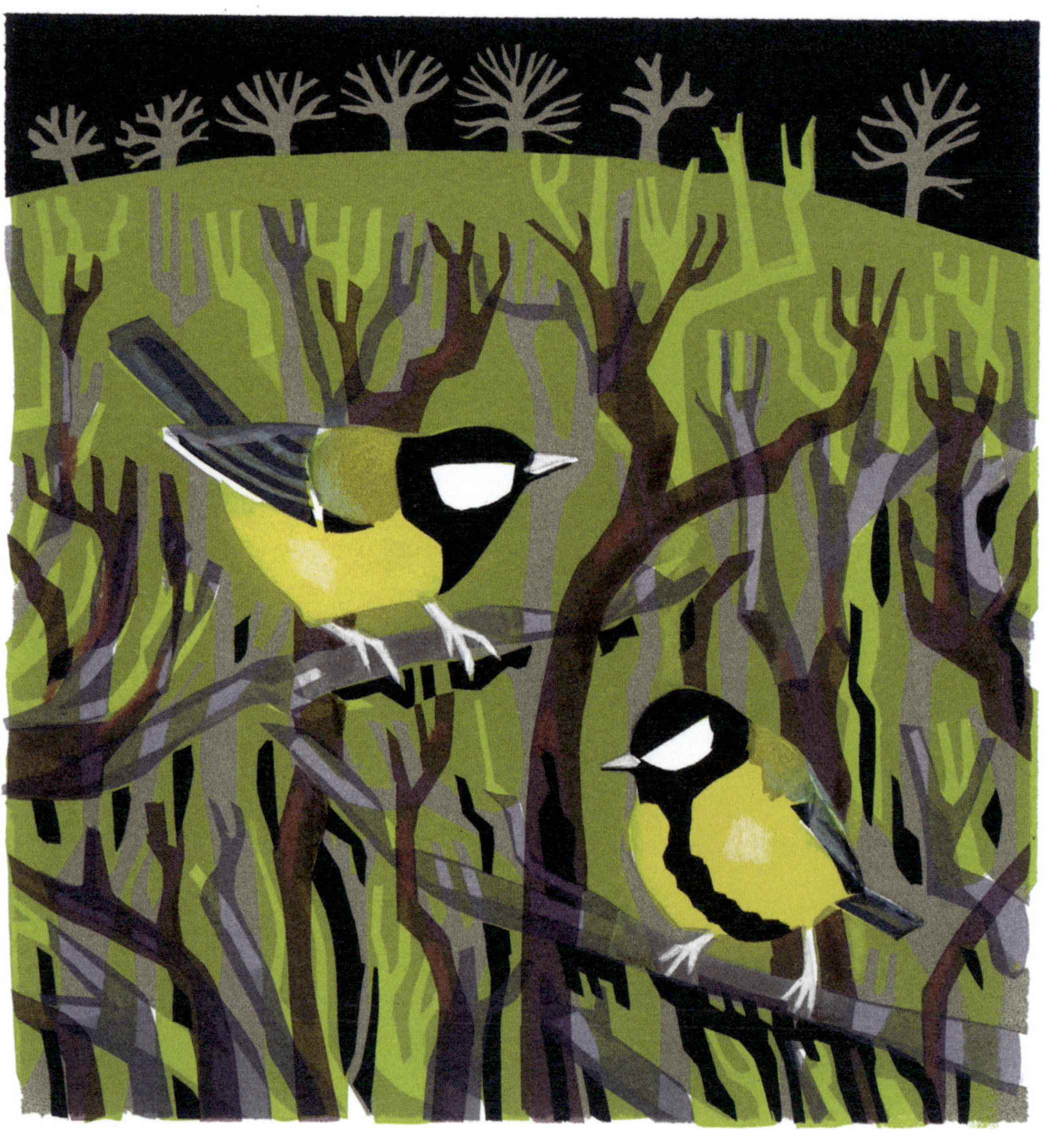

The **great tit** (***Parus major***) anticipates spring with its *dink dink dink*, one of its 40 different calls. This multiplicity is thought to be territorial defence, suggesting an army of great tits awaits an intruder, although each bird knows only a few calls. It means two males can be heard defending adjacent territories with different songs from the repertoire. The brighter the male great tit's breast and the broader its black stripe, the more impressed a female will be. Being big, they are bold hand-feeders. The Wytham Great Tit study (commenced in Wytham Woods, Oxford, 1947) is the world's longest continuous study of an individually marked animal population.

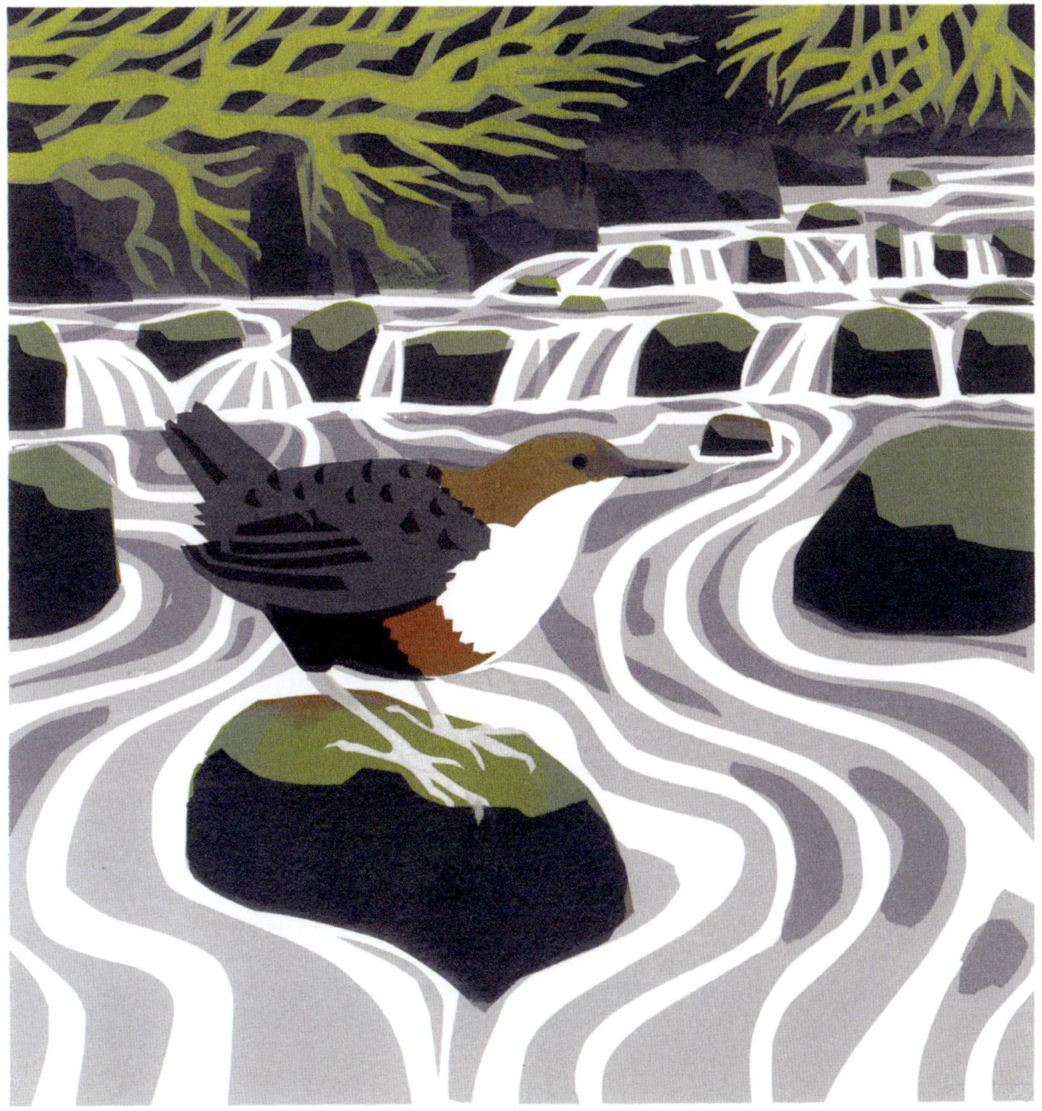

The **dipper** (*Cinclus cinclus*) is the UK's only fully aquatic songbird. It feeds and can escape underwater, using its wings like flippers, feathers waterproofed by an enlarged preen gland. The bones are not hollow like those of most other birds but solid, like the puffin's, to reduce buoyancy. Flaps seal the nostrils, and a high haemoglobin concentration in the blood stores oxygen. The rate at which its body burns energy is one-third lower than in other songbirds. It sings in every season and, surfaced, will bob on a mid-water rock, as if to an applauding audience.

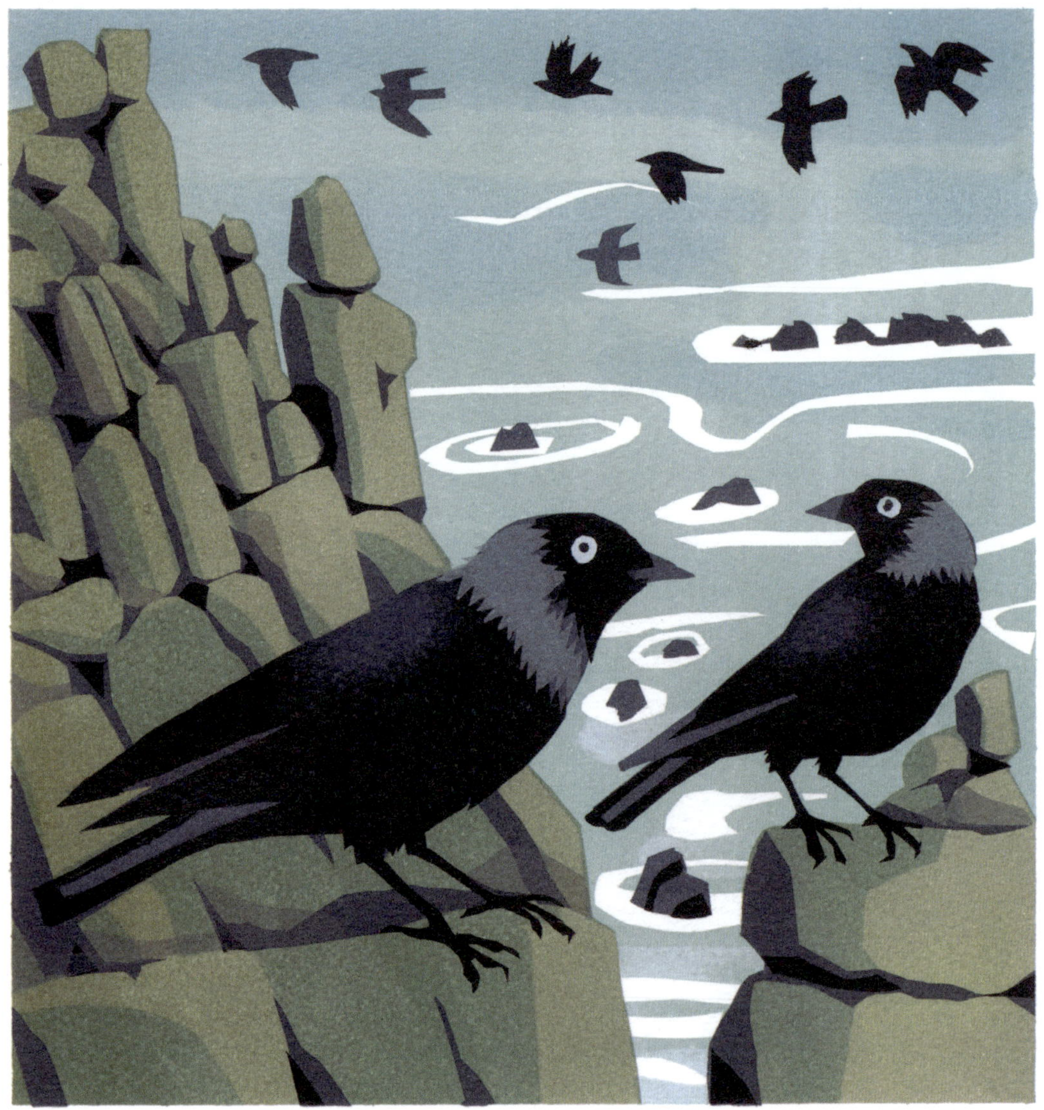

The **jackdaw** (*Coloeus monedula*) nests in rock and wall crevices, tree holes and, notoriously, chimneys. Jackdaws were once a popular pet, as old photographs show. A *County Life* reader recalled one country gentleman who used to take his jackdaw for walks; the bird perched on his shoulder and fed from a trouser pocket. When the owner felt the need to relieve himself, there was a painful misunderstanding. They can form large flocks which, when disturbed, heckle and cackle like struck snooker balls. *Kafka* is Czech for 'jackdaw'. Residents of Malmesbury are called 'jackdaws' after the town emblem.

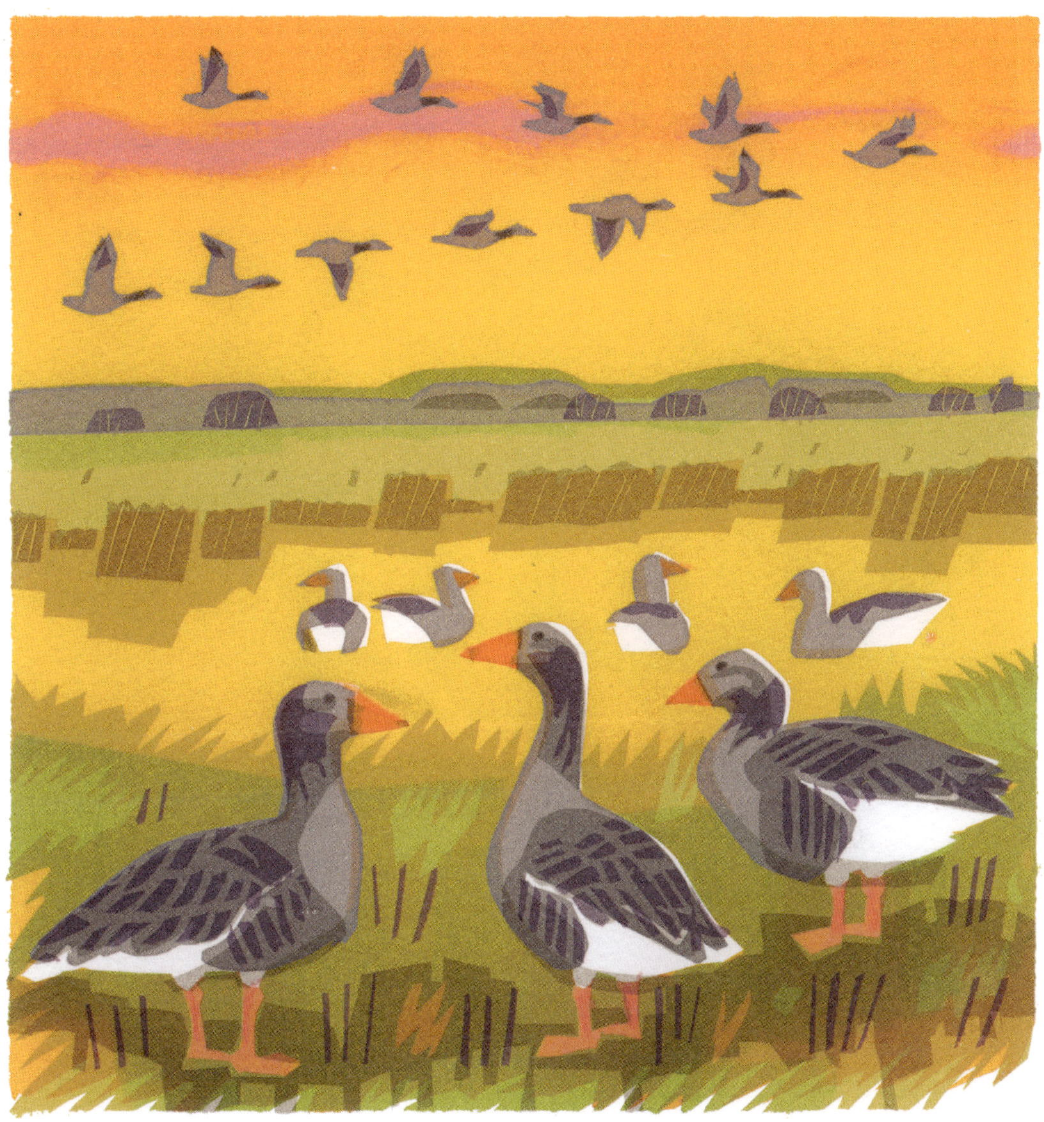

Lag is an old English word for 'goose'. Most farmyard geese derive from the **greylag** (***Anser anser***), domesticated in ancient Egypt, where it symbolised the sun god Ra. Zoologist Konrad Lorenz (1903–89) used greylags to document 'imprinting' for the first time, demonstrating that the offspring of certain species will sometimes follow the first moving object they see. He chose his favourite gosling from a greylag brood but discovered geese are inherently collective. He had to run ahead of all the goslings when coaxing them to fly. In Orkney, greylag numbers have become such a menace to farmland that farmers can kill them legally throughout the year.

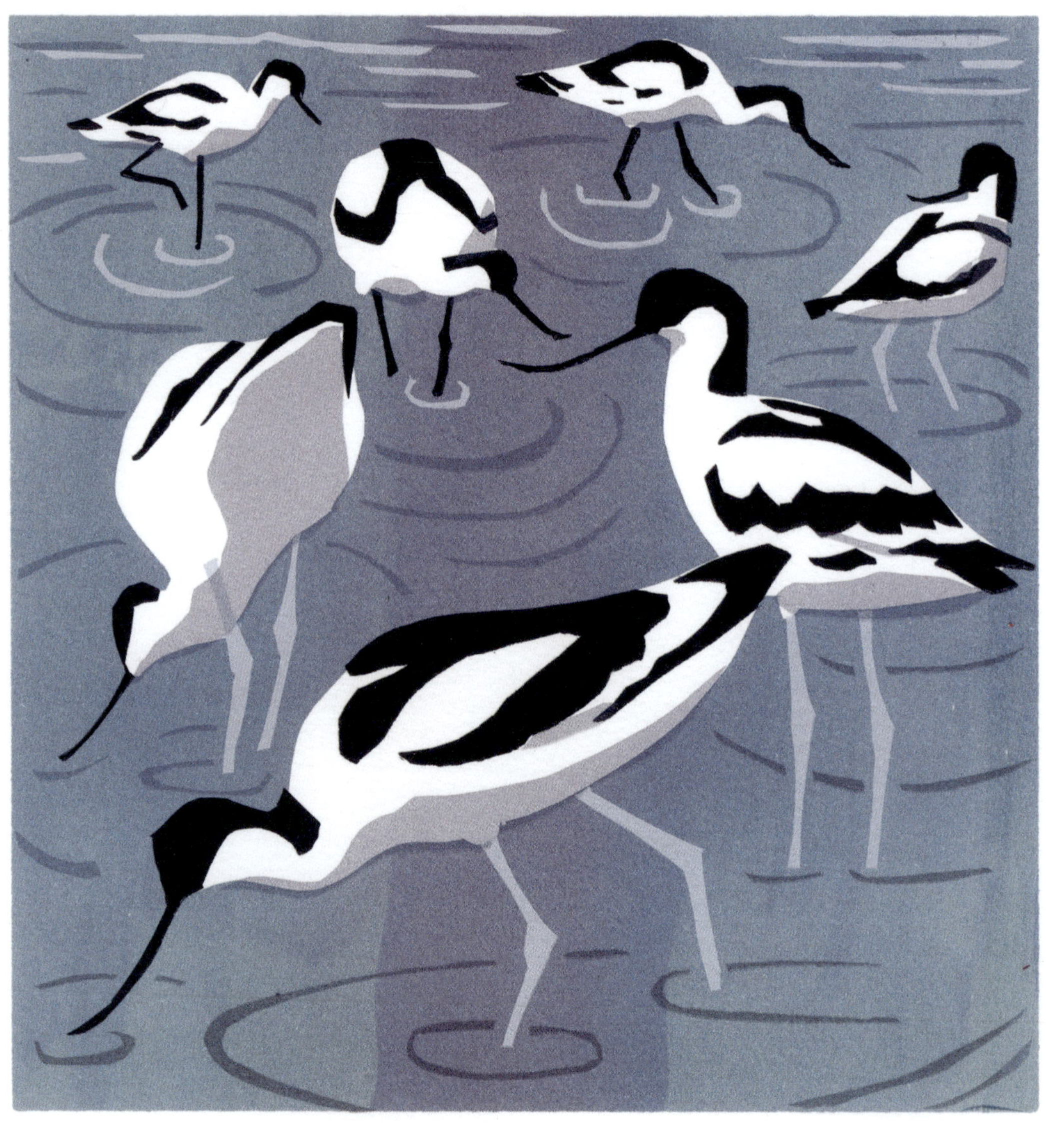

After 100 years of UK extinction the **avocet** (*Recurvirostra avosetta*) returned during the Second World War. By 1947 eight top-secret pairs were nesting at two of the RSPB's first nature reserves, Minsmere and Havergate Island in Suffolk. So synonymous was the Avocet with the RSPB's mission that the organisation chose it for its logo, designed by artist Robert Gillmor (1938–2022) in 1955. There are 1,000 resident pairs in southern England, increased by 8,000 winter migrants. Lieutenant Colonel J. K. Stanford (1892–1971), early volunteer guardian, wrote of his experience in *Bledgrave Hall* that he knew no bird fiercer in defence of its young, a trait latterly earning it the nickname 'Exocet'.

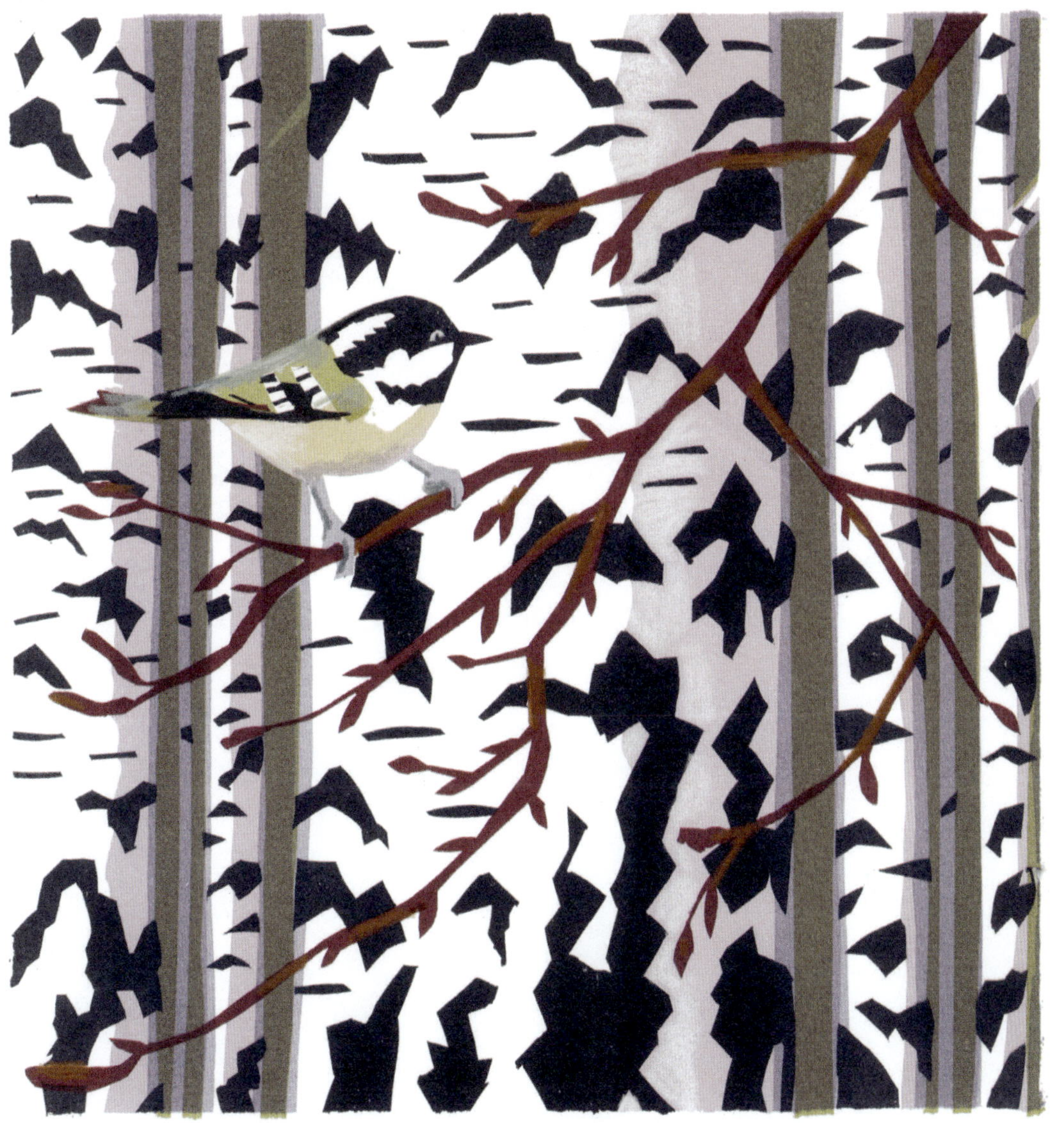

No contact with a wild bird is more magical or consoling than to have one feed from the hand, a praiseworthy achievement in the case of the shy **coal tit** (*Periparus ater*), smallest of the British tits, weighing only 9g. 'Coal' refers to wood charcoal, not mined coal. In medieval Britain the coal tit was called 'little monk', its white nape likened to a tonsure – a bald patch self-inflicted in the West by Christian holy men, especially monks and friars, as a defence against vanity. The practice was abolished for Catholic clergy by Pope Paul VI in 1972.

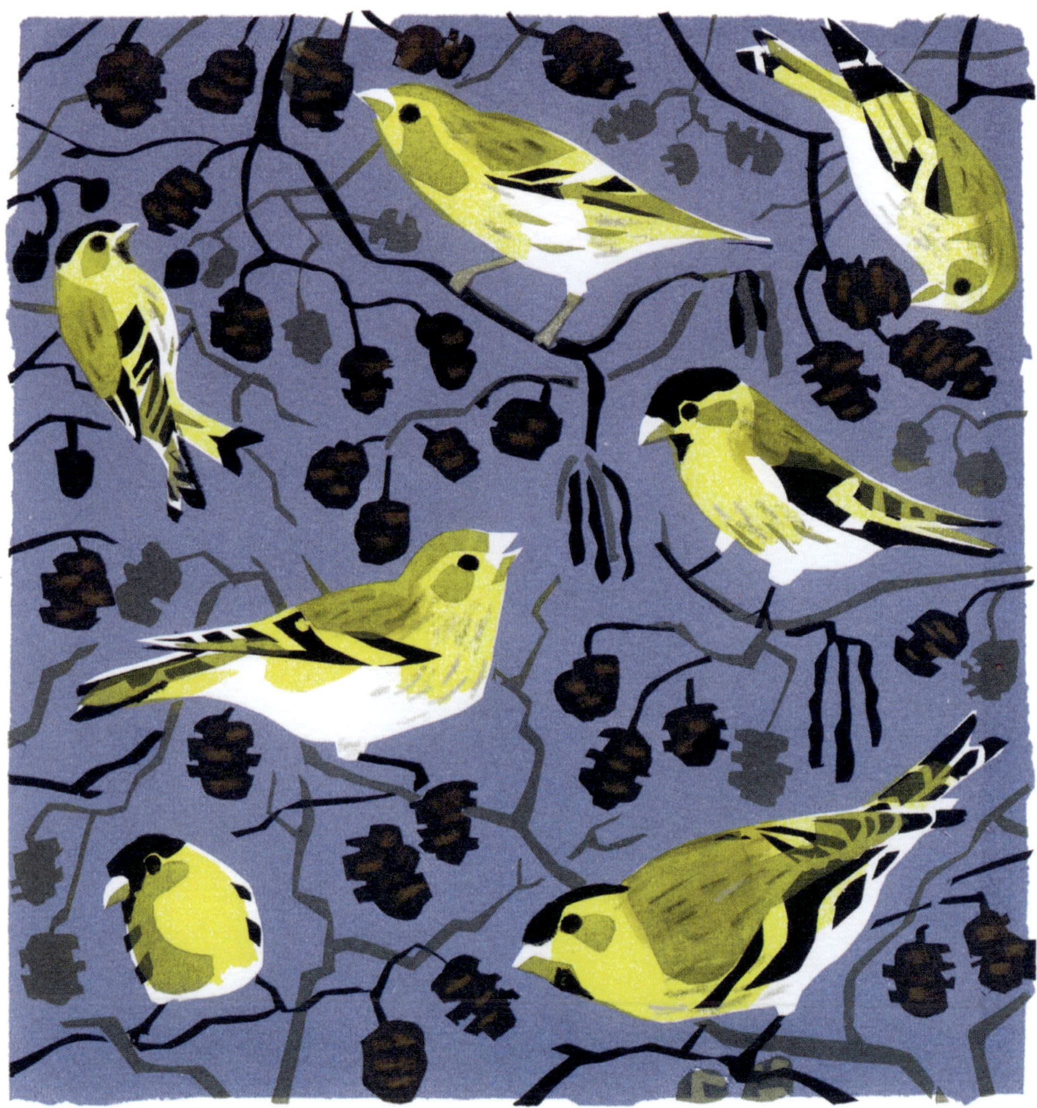

In living memory the **siskin** (*Spinus spinus*) was confined to the Caledonian Forest in the Scottish Highlands. Latterly, continuing commercial forestation in the uplands means that half the United Kingdom's conifer forests are in Scotland, and a large proportion of these are Sitka spruce. Siskins thrive on Sitka spruce seeds, whose cones need warmth to open. When forced by cold weather to wander, siskins feed on alder seeds. The 1960s beginning of the conifer boom coincided with the marketing of garden-bird food. In 60 years, 40,000 UK siskin pairs have become 445,000 pairs, ensuring the species retains a 'green' conservation status.

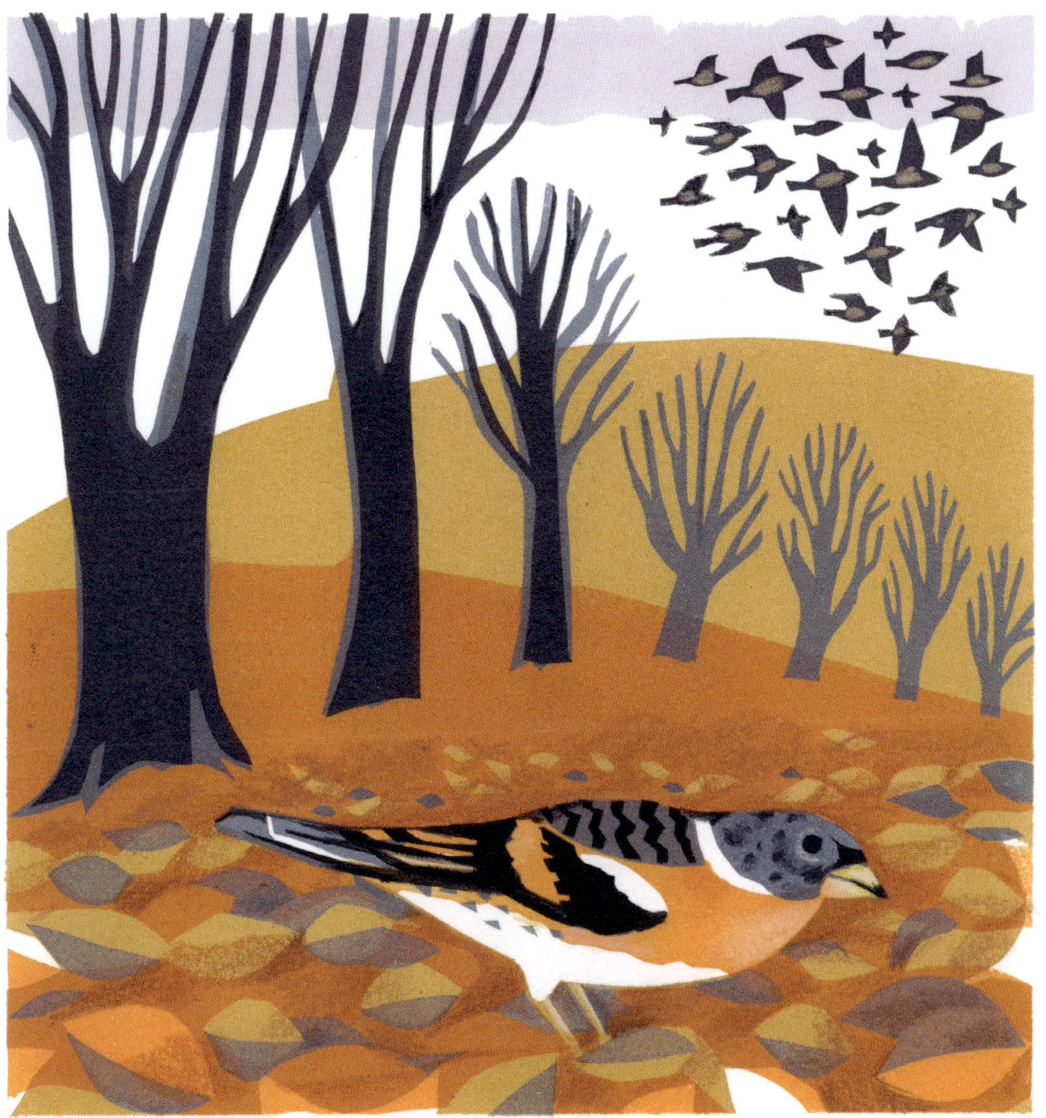

The **brambling** (*Fringilla montifringilla*) is the equivalent of the chaffinch in its northern homeland, Scandinavia and Russia. It is a winter migrant to Britain (up to 1.8m), found wherever there are beech trees, beech-mast its principal food. Accordingly its name is not particularly associated with brambles or blackberries and probably derives from 'brandling', an old word for a creature with a brindled (russet and striped) appearance. Weather dependence makes it the 'forgotten finch': some winters, it is scarce, yet the invasion of the beech woods at Hünibach, Switzerland, by 70m bramblings in 1951/2 remains the largest officially verified avian gathering.

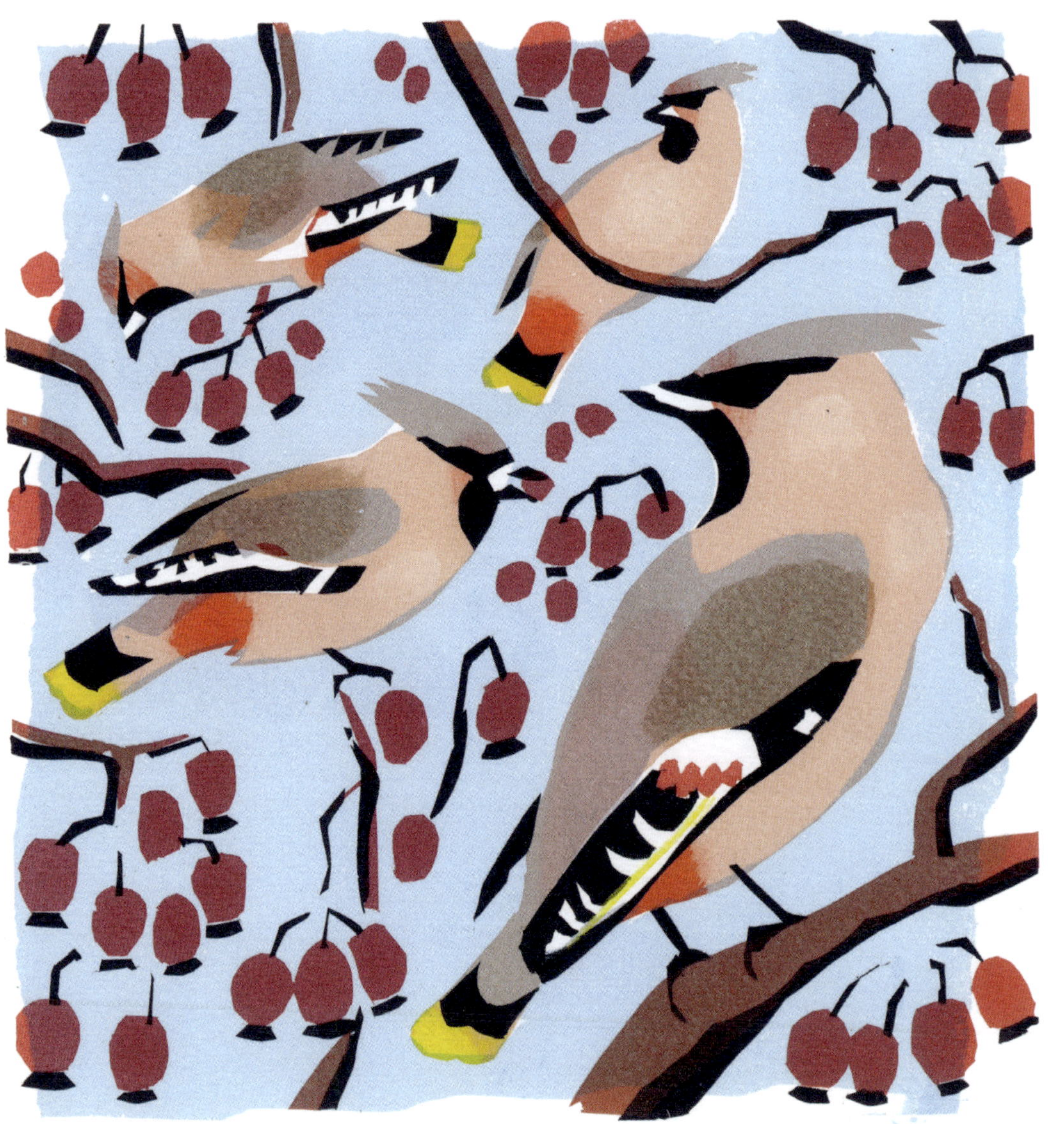

'Migration' is an annual routine; 'irruption' is leaving to survive. The **waxwing** (*Bombycilla garrulus*), winter migrant from the brambling's northern homeland, is subject to irruptions, and up to 10,000 birds have been known to arrive here in wandering flocks seeking sustenance. They guzzle berries so busily in town and out, they can be approached close enough to see the sealing-wax-red points on the secondary wing feathers. 'Waxwing winters' place bird lovers on red alert. A reported flock in Camden settled in a garden. Binoculars! Help!

Bins on,
Bird gone.

Katharine Reid, 'Bird Watching'

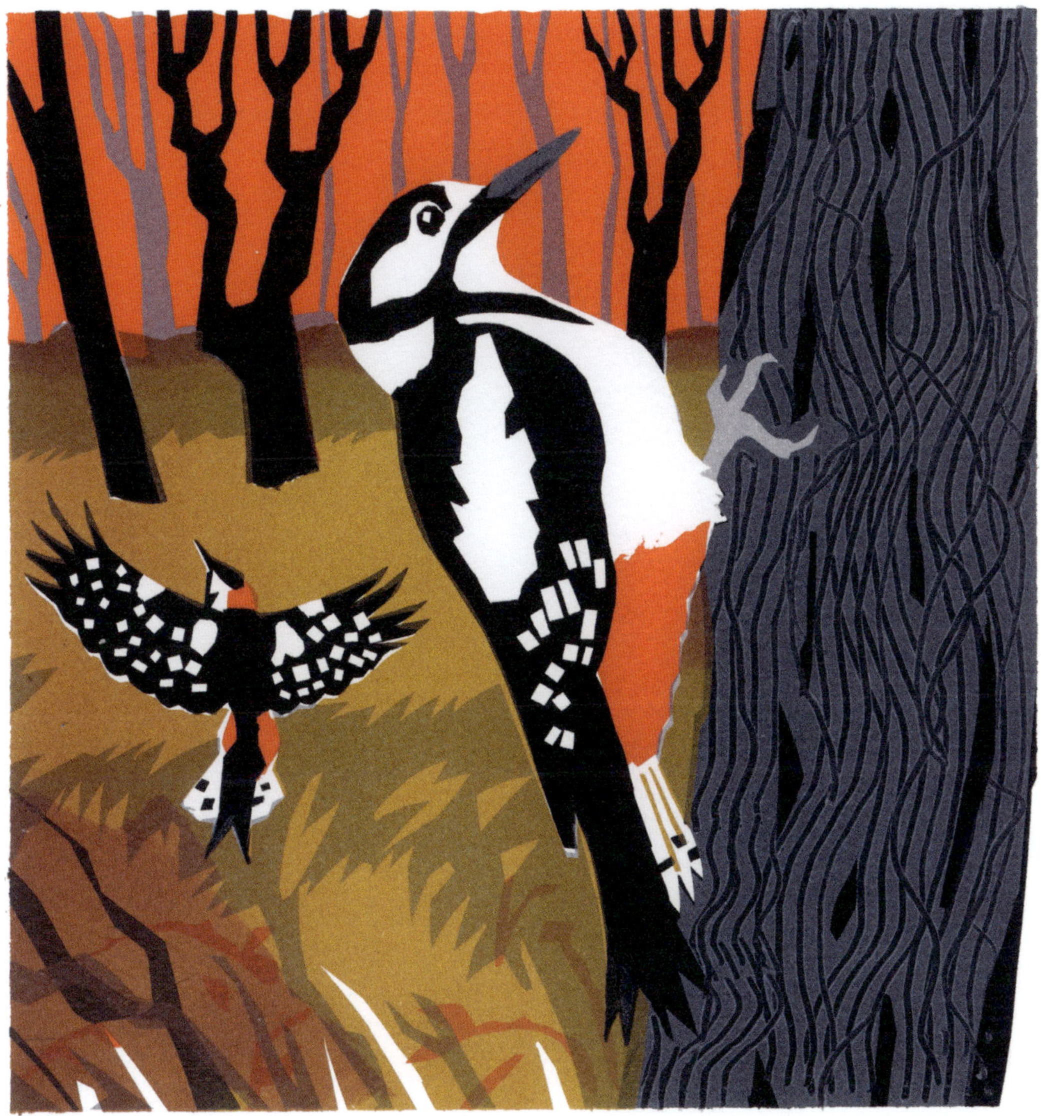

The woodland drumming in spring of the **great spotted woodpecker** (*Dendrocopos major*) is as thrilling to hear as its exotic plumage to see, a taste for peanuts making it regularly visible in country and town gardens. Drumming volume depends on rapidity, the beaks of both sexes striking a branch 5–20 times per half-second. Shock-absorbing tissue in the skull cushions the impact. In spring, they turn carnivores, bolstering an insect diet with the eggs and chicks of smaller birds, especially tree-hole-nesting species like themselves, such as tits and treecreepers.

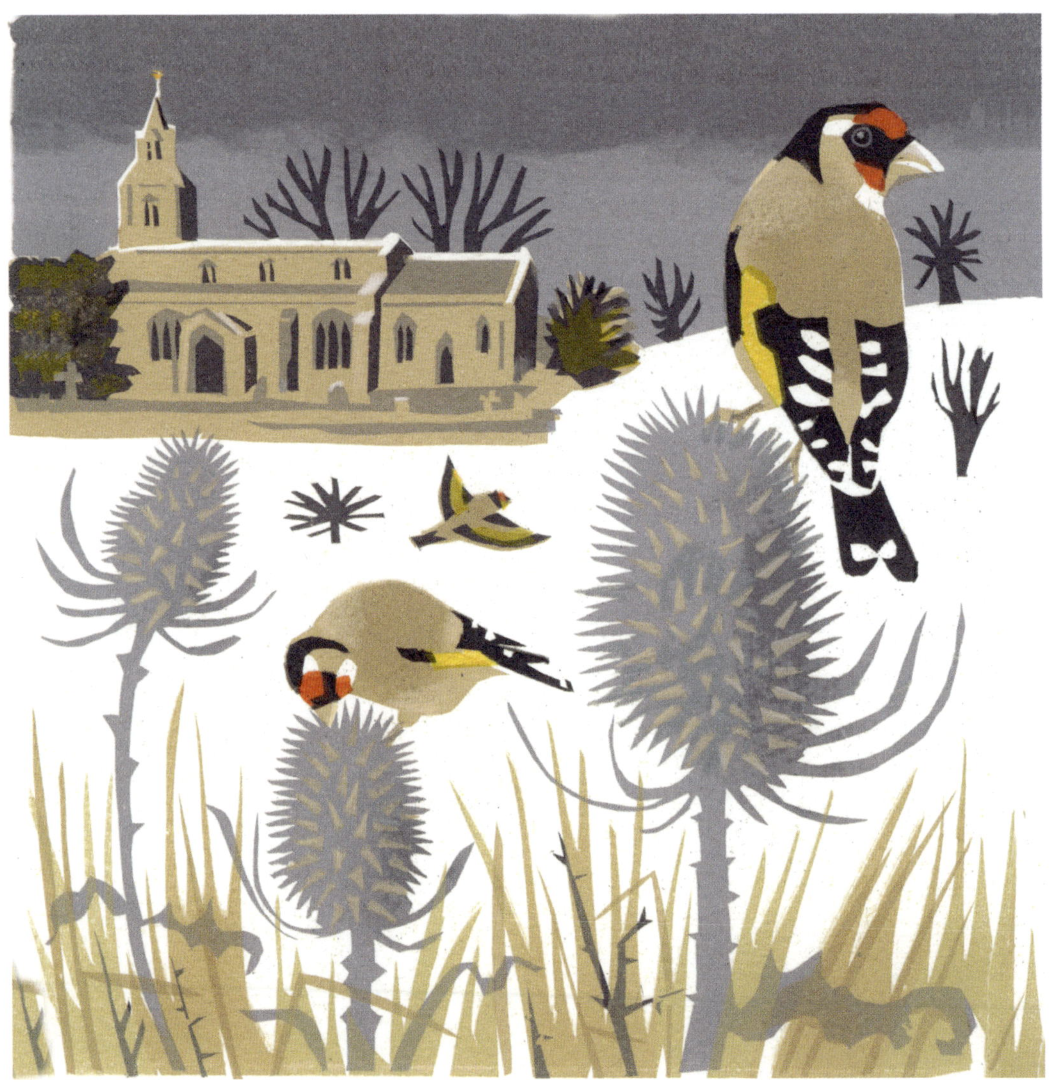

A flock of **goldfinch** (*Carduelis carduelis*, from Latin *carduus*, 'thistle', its favourite food) is called a 'charm', from Old English *c'irm* – after its twittering, not its beautiful appearance. The combination made it a favourite cage bird in Victorian Britain, a trade that, along with featherware, was banned after a campaign by the Plumage League, founded from her Manchester home by Emily Williamson (1855–1936) and later to become the RSPB. The goldfinch's taste for imported African nyjer seed has made its twittering commonplace in towns.

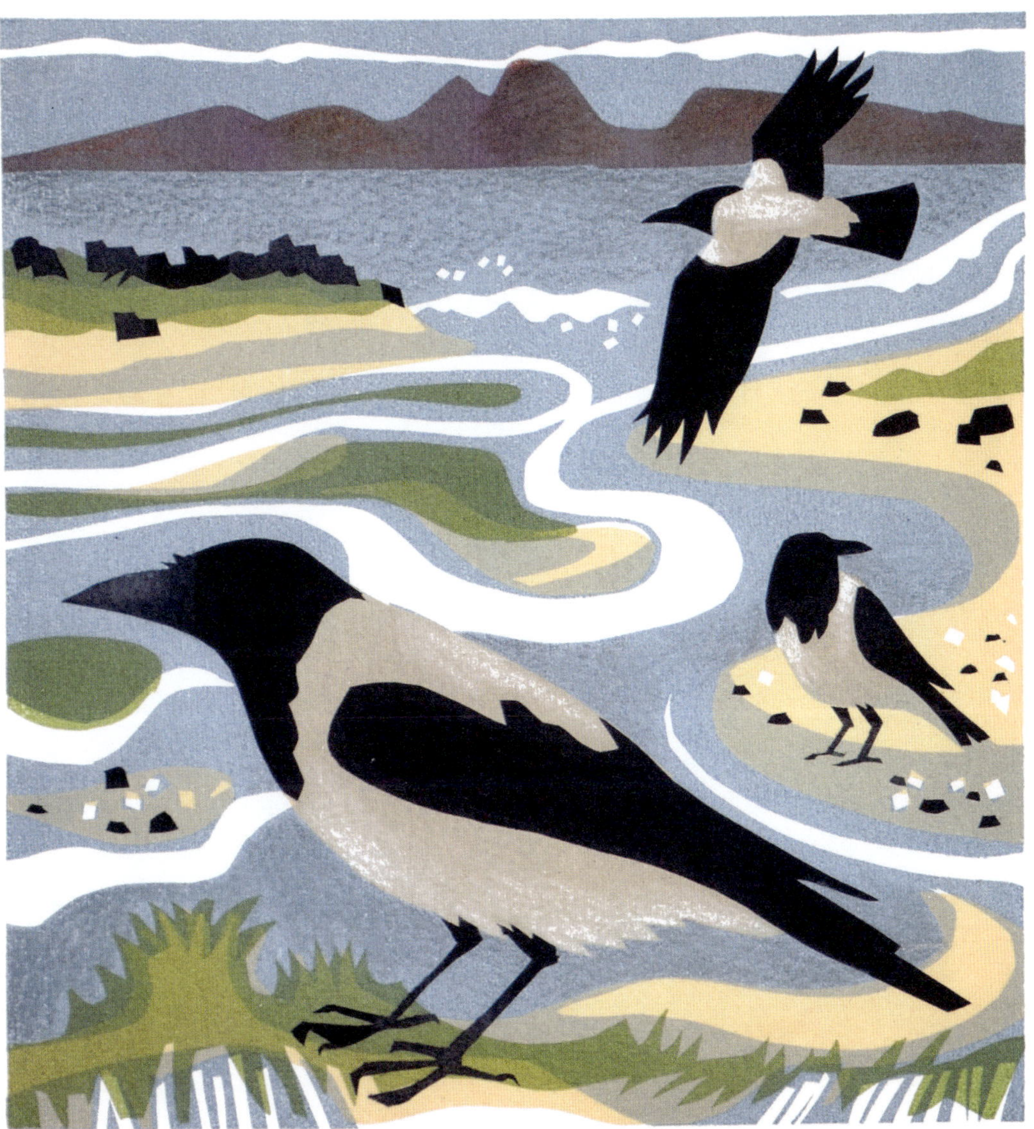

The grey-mantled **hooded crow** (*Corvus cornix*) was once seen as a subspecies of the carrion crow (*Corvus corone*), though it differs in plumage and location; it was reclassified as a separate species in 2002. The UK's 285,000 'hoodie' pairs favour north-west Scotland, the Isle of Man and Northern Ireland; the carrion, at 1.05m territories, is ubiquitous elsewhere. Cross-breeding occurs along the Highland line, the result being hybrid birds with intermediate plumage. On mainland Europe, they divide east and west – carrions in France and Spain, hoodies in Italy and Greece.

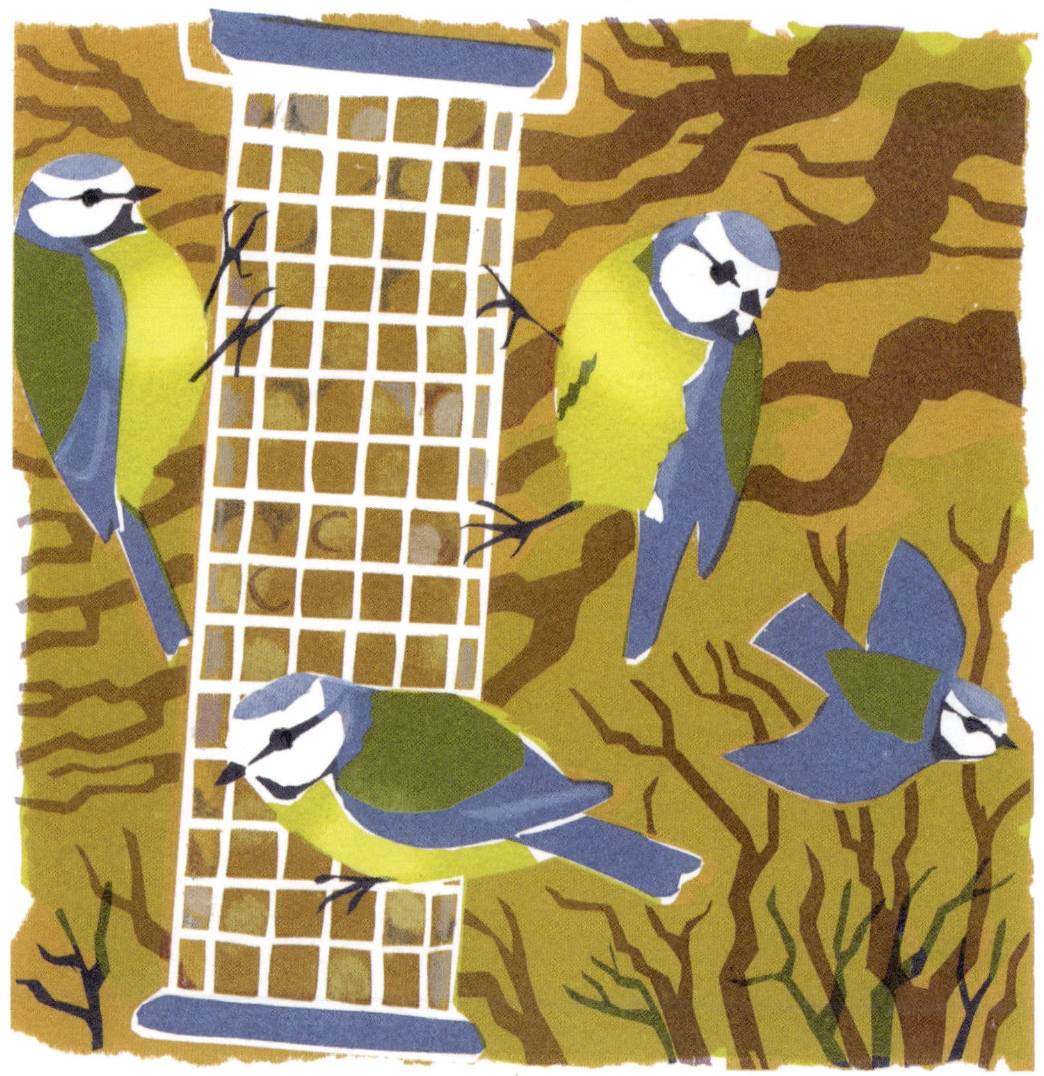

In 1921 at Swaythling, Southampton, a genius **blue tit** (*Cyanistes caeruleus*), to get at the risen cream, pierced the foil cap of a glass milk bottle left on a doorstep by the morning milkman. Blue tits lack enzymes to digest the lactose in milk; cream is almost lactose-free. Other blue tits, great tits, even other bird species followed suit. 'Tit' meaning 'small' in the fourteenth century was attached to 'mose', from Old English *mase*, 'small bird', forming 'titmose', later 'titmouse'.

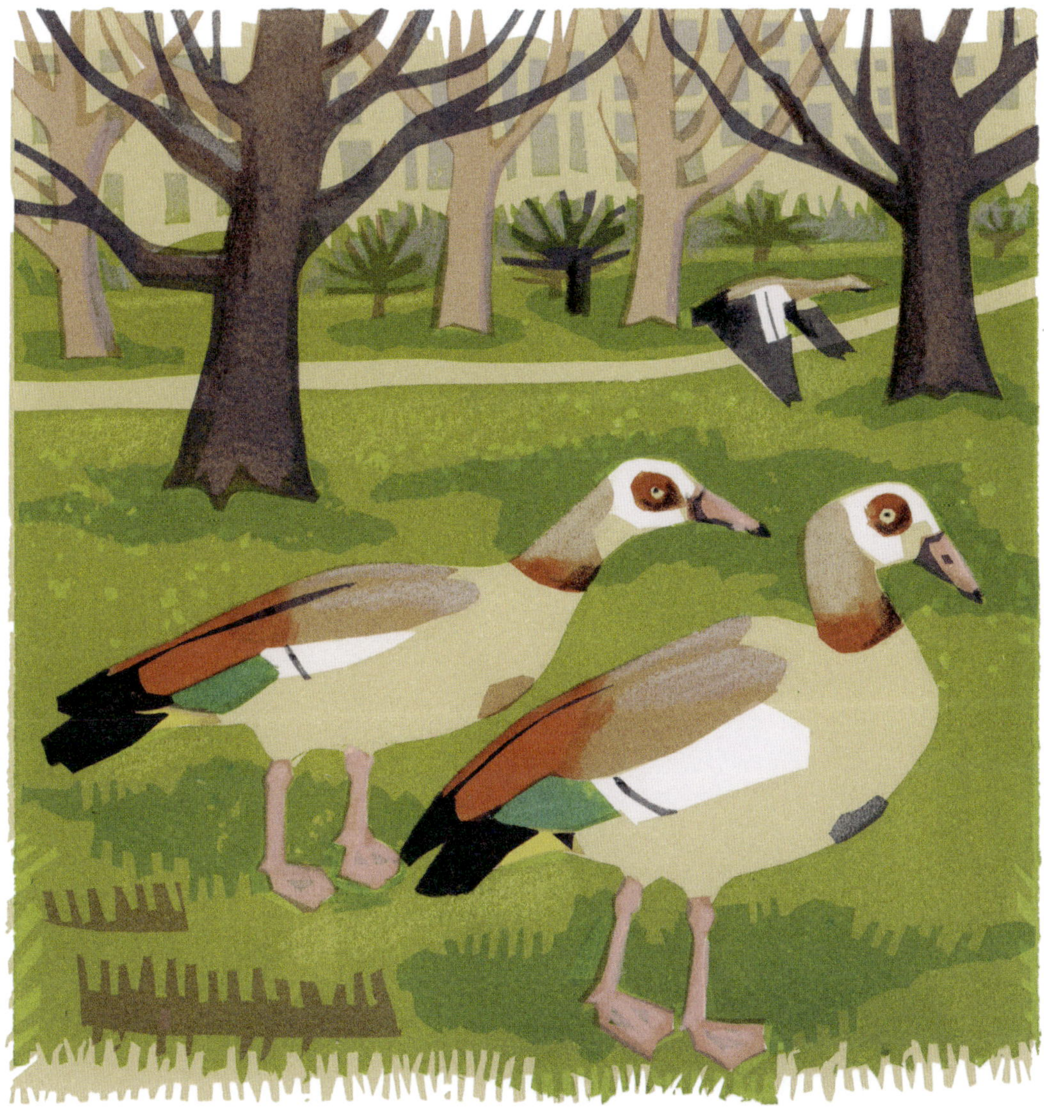

The **Egyptian goose** (*Alopochen aegyptiaca*) sounds husky or, when agitated, like an old banger being hand-cranked. It arrived in the British Isles as part of the ornamental wildfowl collection of King Charles II (1630–85) in St James's Park, London. It nests in tree holes – a shock in St James's Park to see one perched high in a plane tree or wandering the 6m-thick (19.5ft) concrete roof of the World War II bomb-proof Admiralty Citadel on Horse Guards Parade. Selection for picturesque lakes in the landscaped parks of the nobility soon created a wild population, although it was not listed as a British bird until 1971. Egyptian geese can nest late (November) and early (January), the young growing fast to reduce the disadvantage.

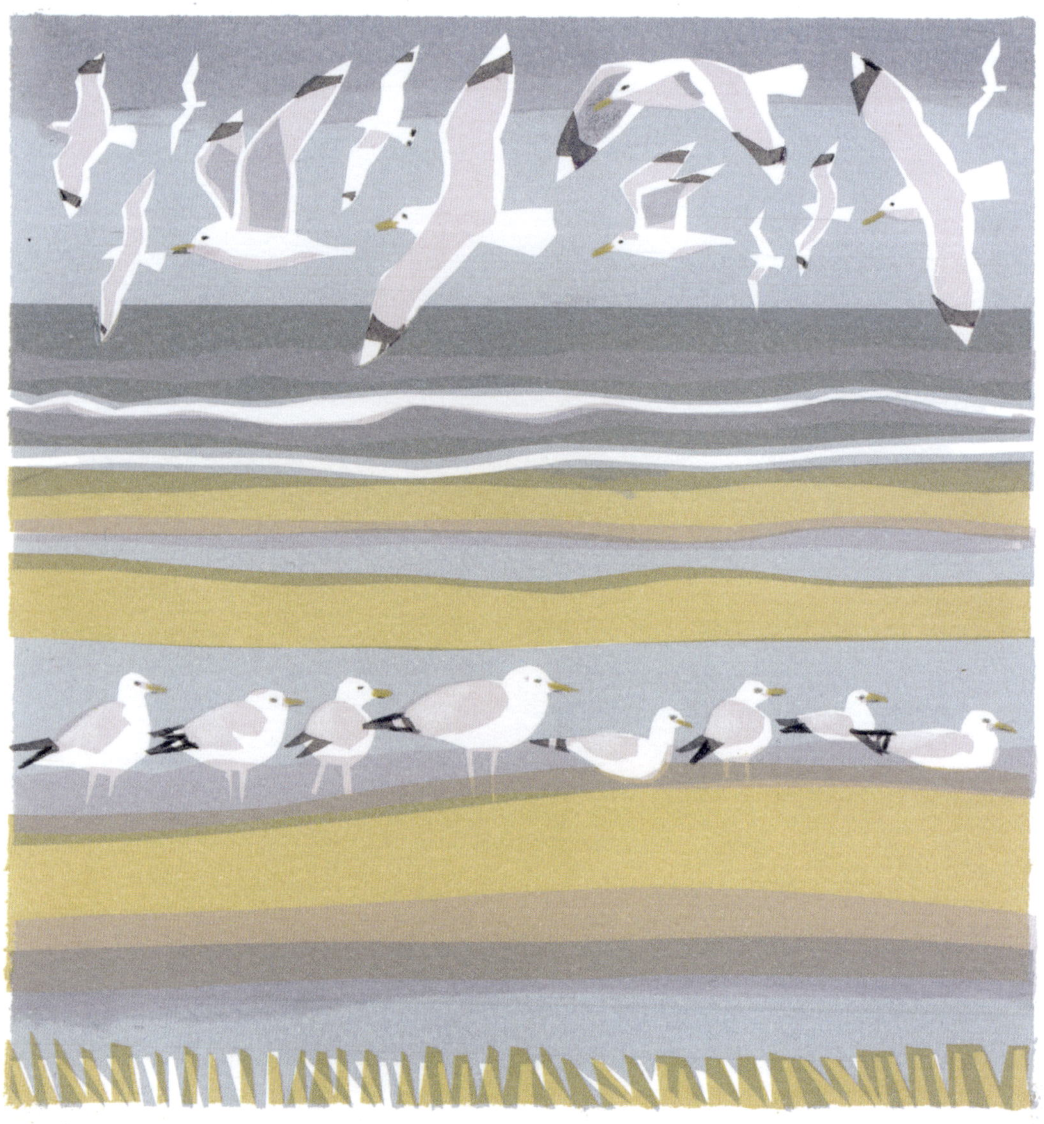

Gulls are challenging to twitchers because, of the baker's dozen species regularly seen in the UK, five appear mostly in winter and then in numbers under 1,000. The **common gull** (*Larus canus*) is fifth most numerous among resident gulls, at 48,000 pairs, and winter adds 600,000 migrants. When residents have finished breeding, mostly in Scotland, they are very similar to the UK's second-commonest breeding gull, the black-headed, which sheds its black head in winter. Seen together, they are easiest told apart by the common's yellow legs and bill, as opposed to the red of the black-headed's. Twitchers will hope to spot a yellow-legged gull, whose annual numbers only reach 840 winter migrants.

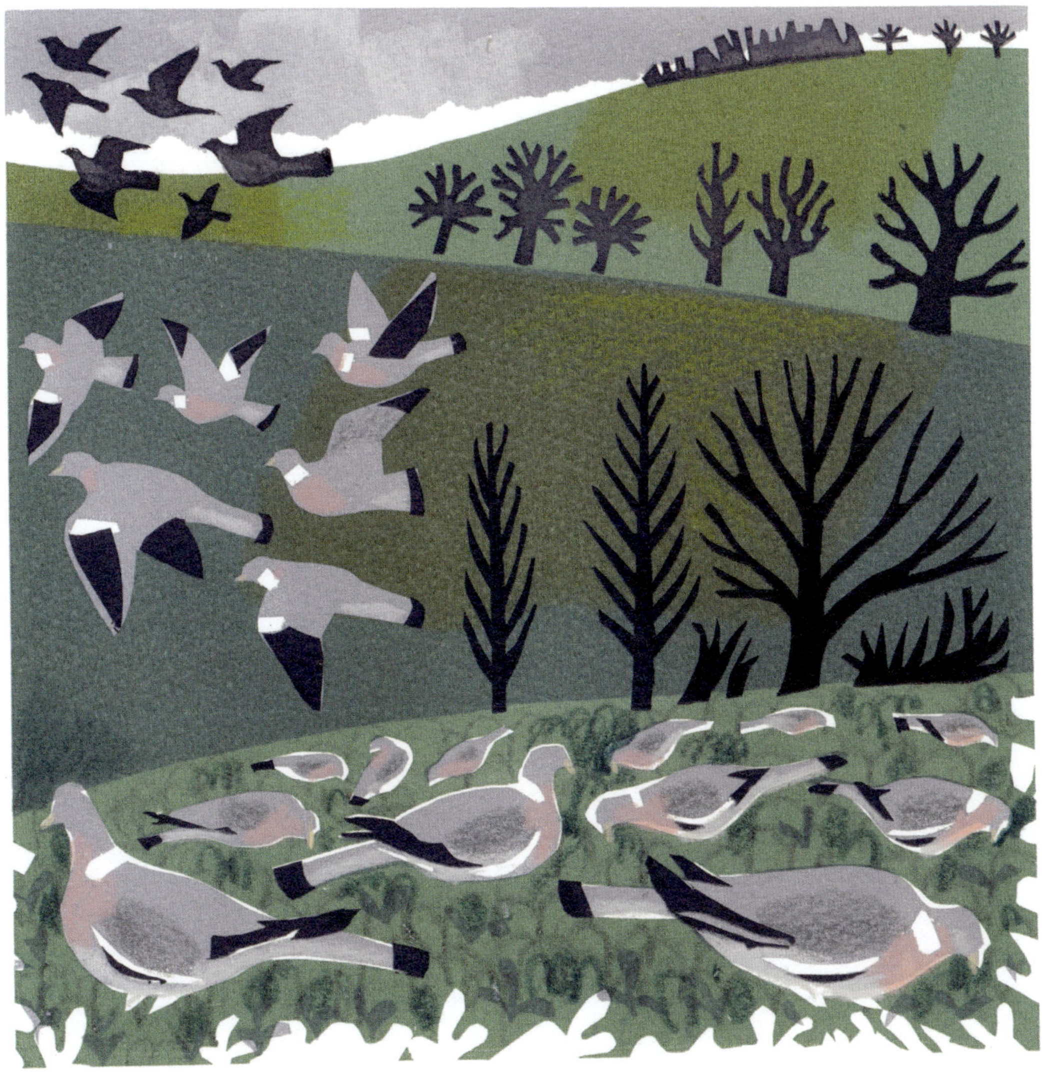

The **woodpigeon** (*Columba palumbus*) was traditionally known as the 'ring dove' – more appropriate to its appearance and soothing, *coo-cooing* call. Native to town and country, it prospers due to milder winters encouraging year-round breeding and produces a secretion called 'crop milk' to feed young, a behaviour also seen in flamingos and male emperor penguins. In 2016, a winter flock of 90,000 was recorded in Gloucestershire. At 5.15m pairs, it is the fourth most populous UK bird. Pigeons and doves (Columbidae) – the extinct dodo was one – suck rather than dip-tilt to drink. During spring and summer dawns, woodpigeons can rhythmically coordinate their cooing over many miles. Tame in town, a fear of predators makes them wary in the wild.

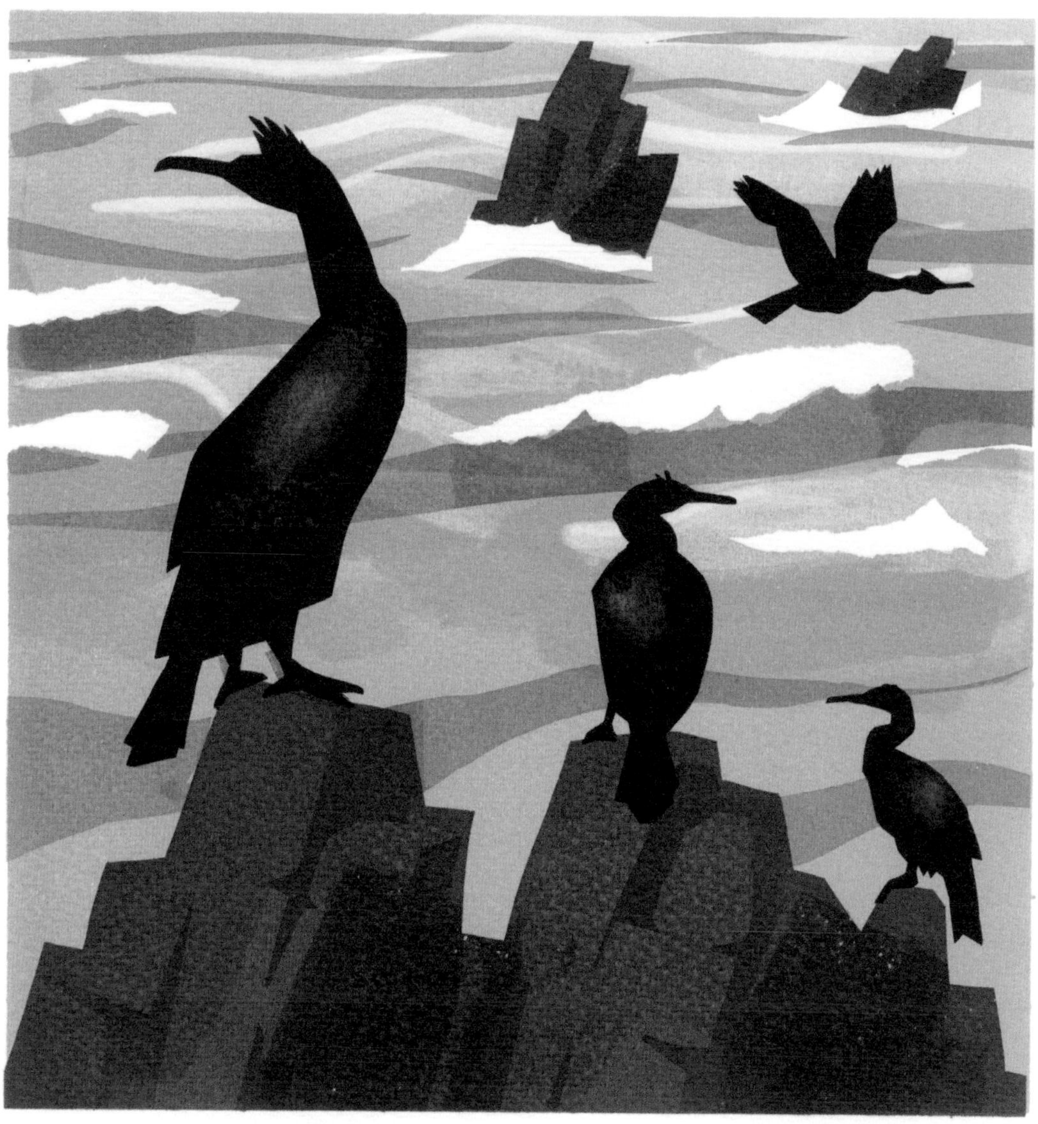

One genus and three species of animal are named after Aristotle (384–322 BC), the father of biology; the only bird species is the **shag** (*Gulosus aristotelis*). Its English name comes from Old Norse for 'beard', after its crest in the breeding season. It is a marine bird, unlike its more visible, inland-inclined, cormorant cousin (8,900 resident pairs; 54,000 winter migrants). The shag is most visible in winter when migrants swell the resident UK population (20,000 pairs; 90,000 winter migrants) to occupy the rocky parts of the British coastline. Shags hate rain, so favour caves. Wartime London restaurants served shag as 'black duck'. Traditionally, before cooking, it was thought best to bury the corpse for a week.

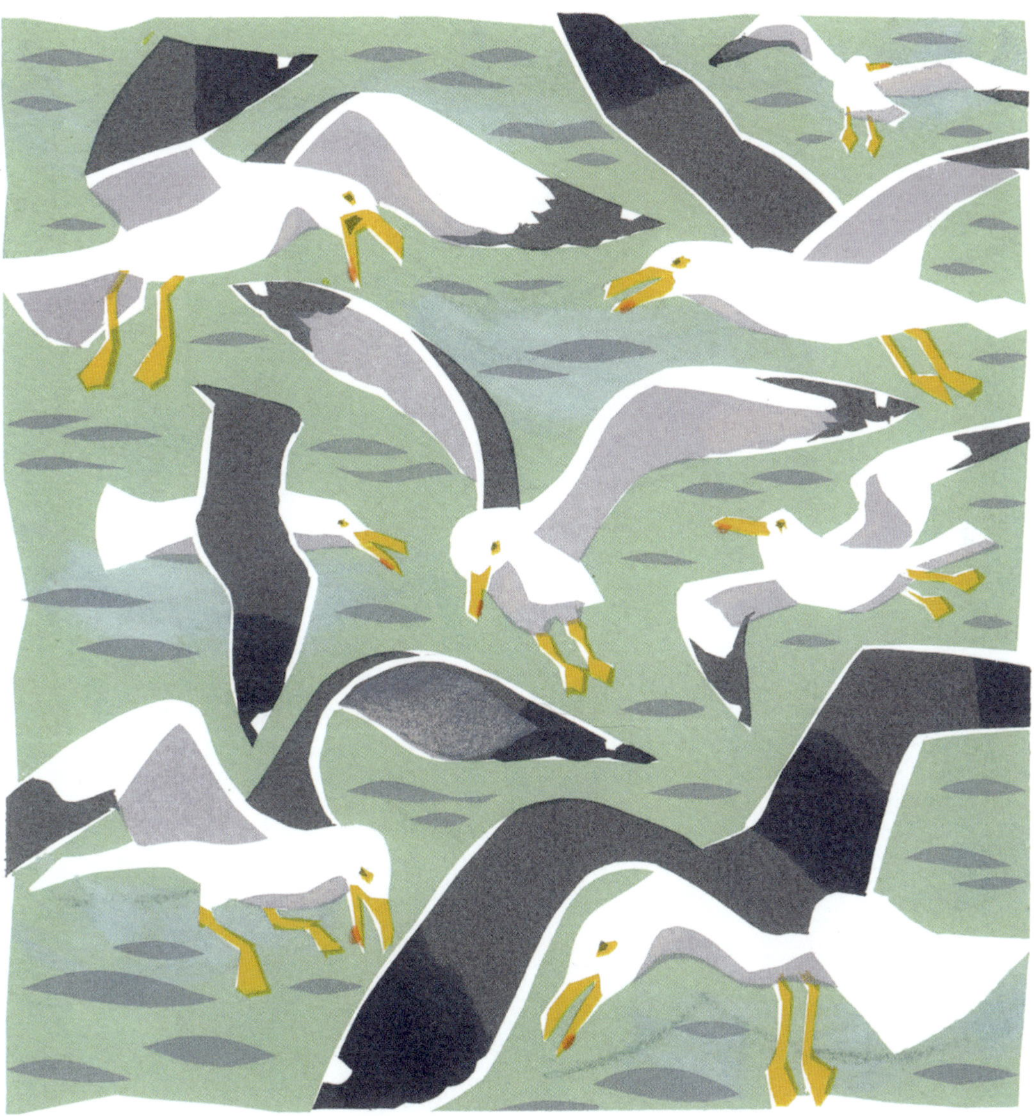

We call gulls 'seagulls' but, like the similarly-sized herring gull, the **lesser black-backed gull** (*Larus fuscus*) has been encouraged by garbage to relocate from sea to town, where it prospers as a scavenging bin raider. It used to be a winter migrant from the coasts of Iberia and North Africa. The first summer breeding was in Merthyr Tydfil in 1945, and there are now 110,000 UK pairs. By nature a sand nester, it has adapted to flat roofs. Like herring gulls, lesser black-backeds prey on chicks of other species, and a couple of individuals in central London became specialist hunters of adult feral pigeons, grabbing one from a feeding horde and drowning it in a nearby lake.

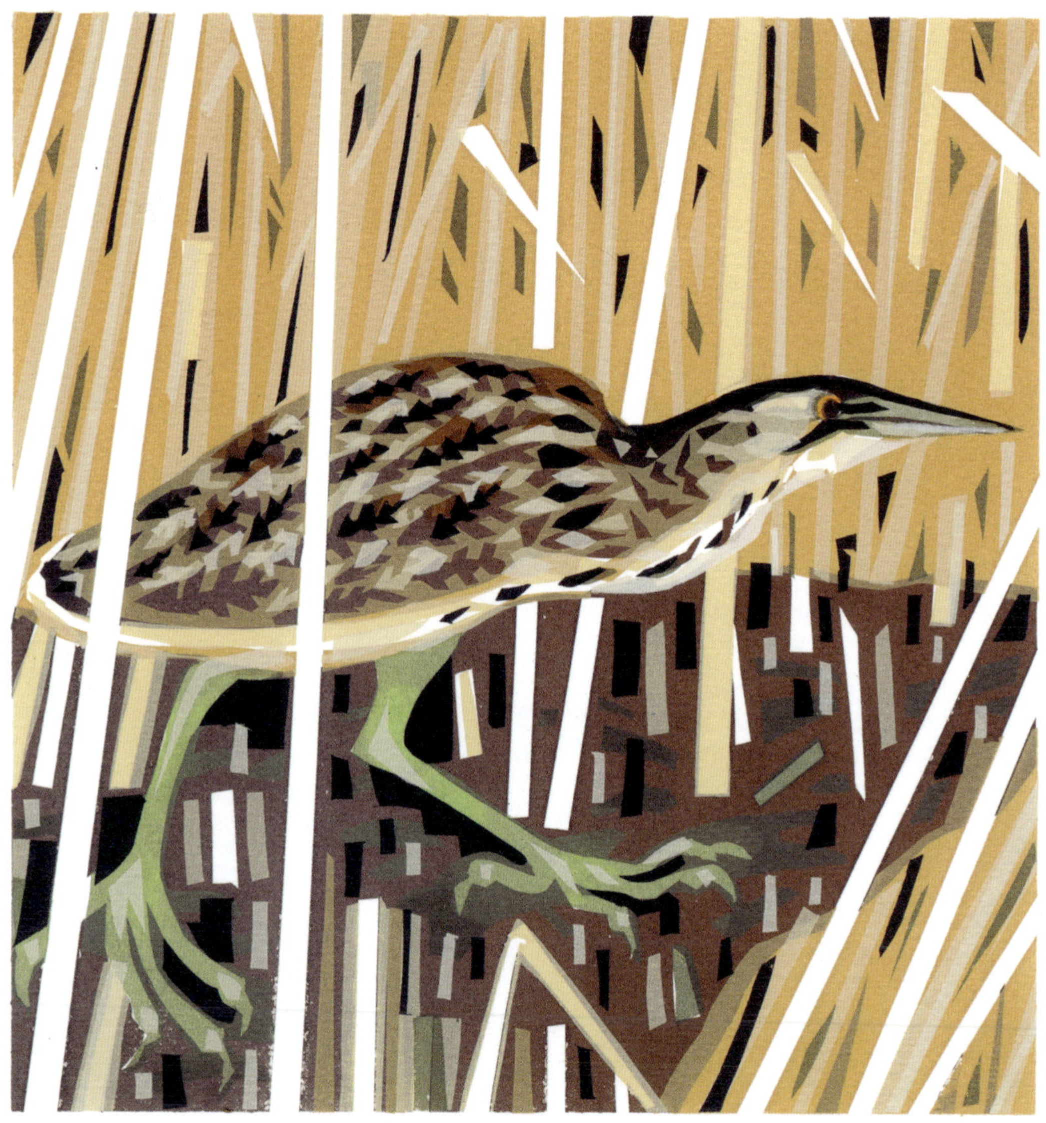

In the words of David Lodge, self-described 'smitten-by-the-bittern man of the marsh', especially Stodmarsh Nature Reserve in Kent, the **bittern** (*Botaurus stellaris*) is 'an extremely frustrating bird to study'. The male's booming mating call (*Botaurus*, 'bull's bellow') makes it an avian celebrity, though even with winter migrants more than doubling the resident UK population, there are under 1,000. The boom can sound like a foghorn or be as faint as a gentle blow into the empty neck of a bottle. Of the heron family, a bittern rarely emerges from its reed-bed habitat. Alarmed, it can elongate itself to look like a cluster of reeds. Carry startled one into camouflage mode on a path. It flew before she could take a photograph.

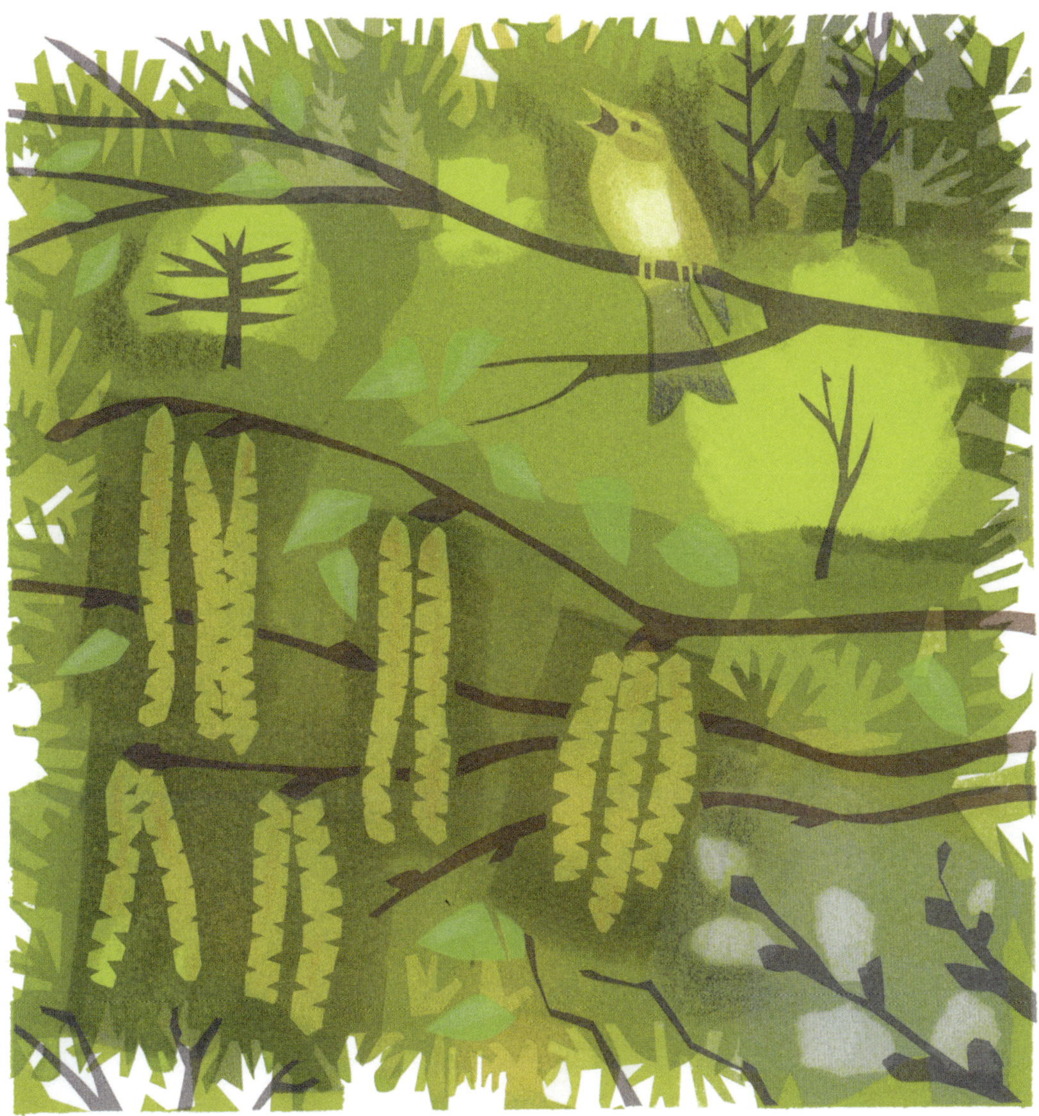

Numbering more than 2m, the most numerous of the UK's summer warblers from Africa is the **willow warbler** (*Phylloscopus trochilus*). Song is the surest means of identifying warblers, especially the scrub-dwelling, similarly plumaged, well-named 'leaf' variety, such as the willow warbler. Henry Douglas-Home described its wistful 20-second cadence in his avian memoir *The Birdman* as 'finer than all the stronger voices it contends with'. The bird was first classified in 1789 by naturalist Rev. Gilbert White (1720–93), who was also first to denote a kiss with an x.

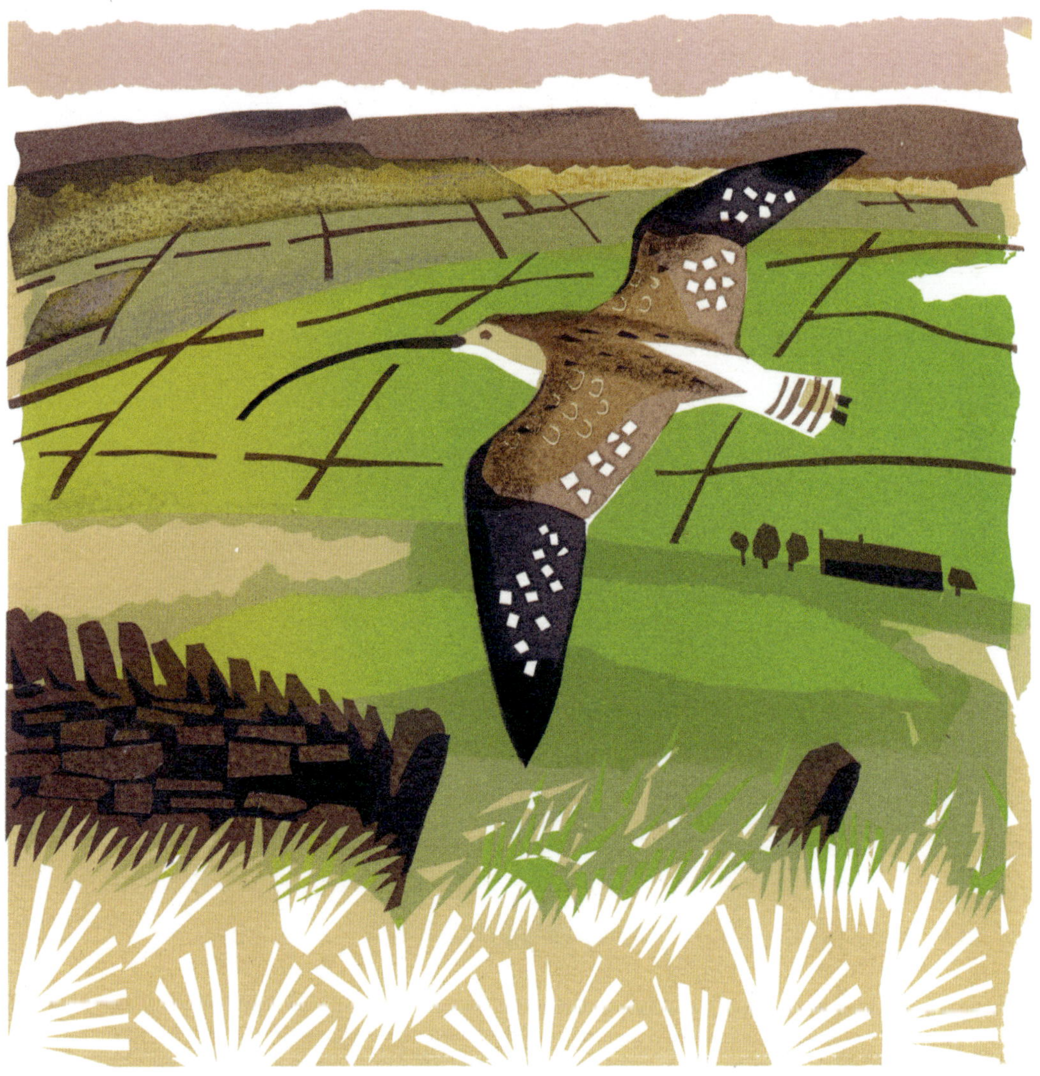

The curve-billed **curlew** (*Numenius arquata*, 'arched like a bow'), or 'whaup' in Scotland, is Europe's largest wader. Its call, rising to a bubbling crescendo, is particularly haunting when heard in its wilderness breeding grounds. Robert Louis Stevenson (1850–94) includes the curlew in 'To S. R. Crocket', his sonnet to his Ayrshire novelist friend, associating it with the 'martyrs' – Presbyterian covenanters, who sought refuge on the moors, as birds do, for safety.

> *Where about the graves of the martyrs the whaups are crying,*
> *My heart remembers how!*

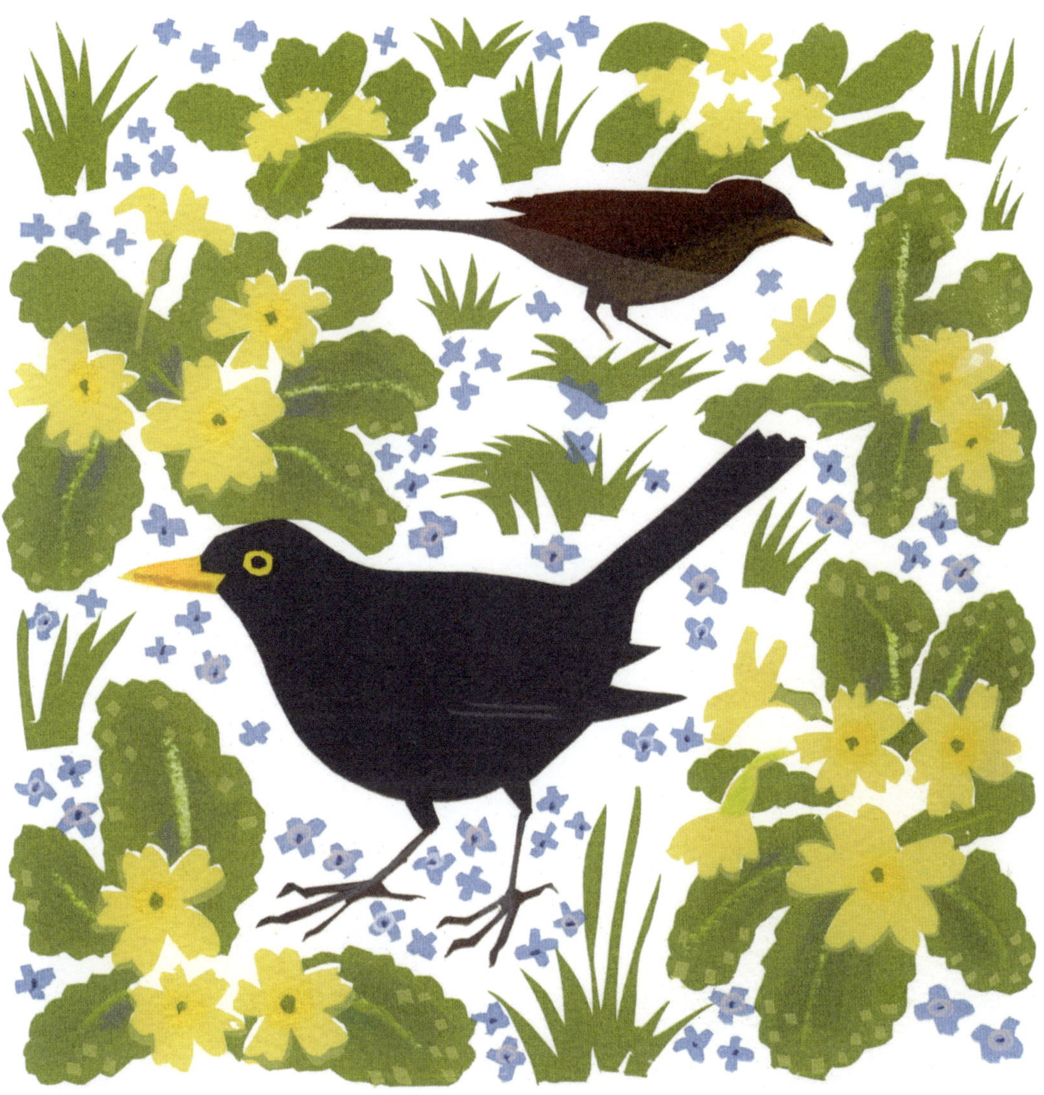

What Pliny the Elder (d.79 AD) wrote of the nightingale applies to several songbirds, notably the **blackbird** (*Turdus merula*): 'Nay, they strive who can do best, and one laboureth to excel another in varietie of song and long continuance' (*Naturalis Historia*, 77AD). Their *pink-pinking* alarm call is encouraged in winter by the threatening arrival of European migrants. A low, liquid *pok* can also punctuate snow-dumbed days. Artist and author Denys Watkins-Pitchford (1905–1990), pseudonym BB, described the male's plumage as the densest black of any bird, its orange beak the perfect contrast.

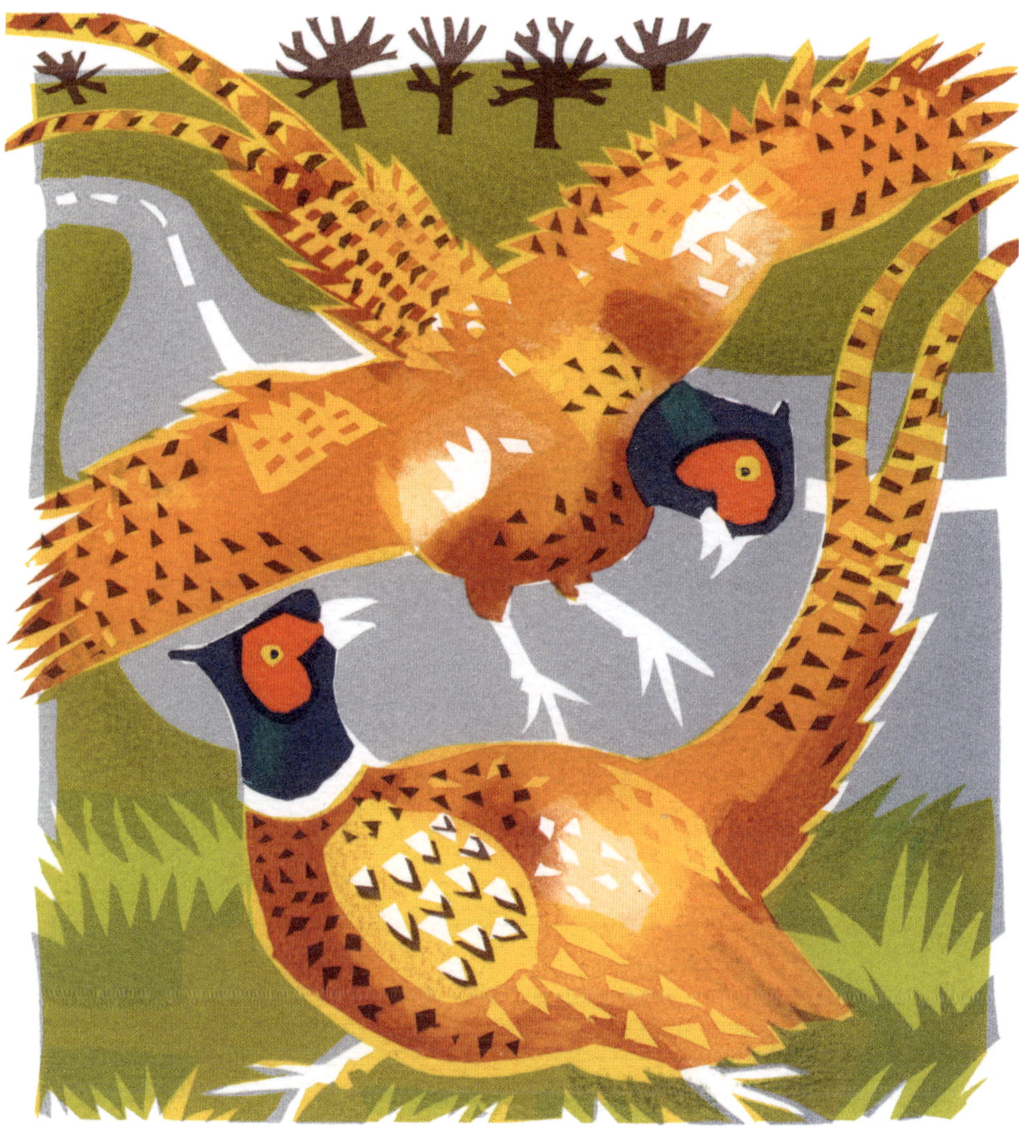

The **pheasant** (*Phasianus colchicus*, meaning 'from Colchis', now Georgia) is polygamous, like 5 per cent of birds, with hens choosing a mate based on spring cock-fights. The Romans probably brought the pheasant to Britain, and the nineteenth-century shotgun introduced driven pheasant shoots. Today, commercial shoots supplied with farmed pheasants (43m in 2019) are big business, contributing £2 billion p.a. to the rural economy. Shooting counteracts intensive farming – pheasants need woods and hedgerows, beneficial for flora and fauna. In spring, pheasant country resounds to cocks crowing accompanied by wing-beating flurries. They also crow at the jolt of sounds, which they can hear in advance of humans, and to announce winter ascents to roost in wind-breaking conifers.

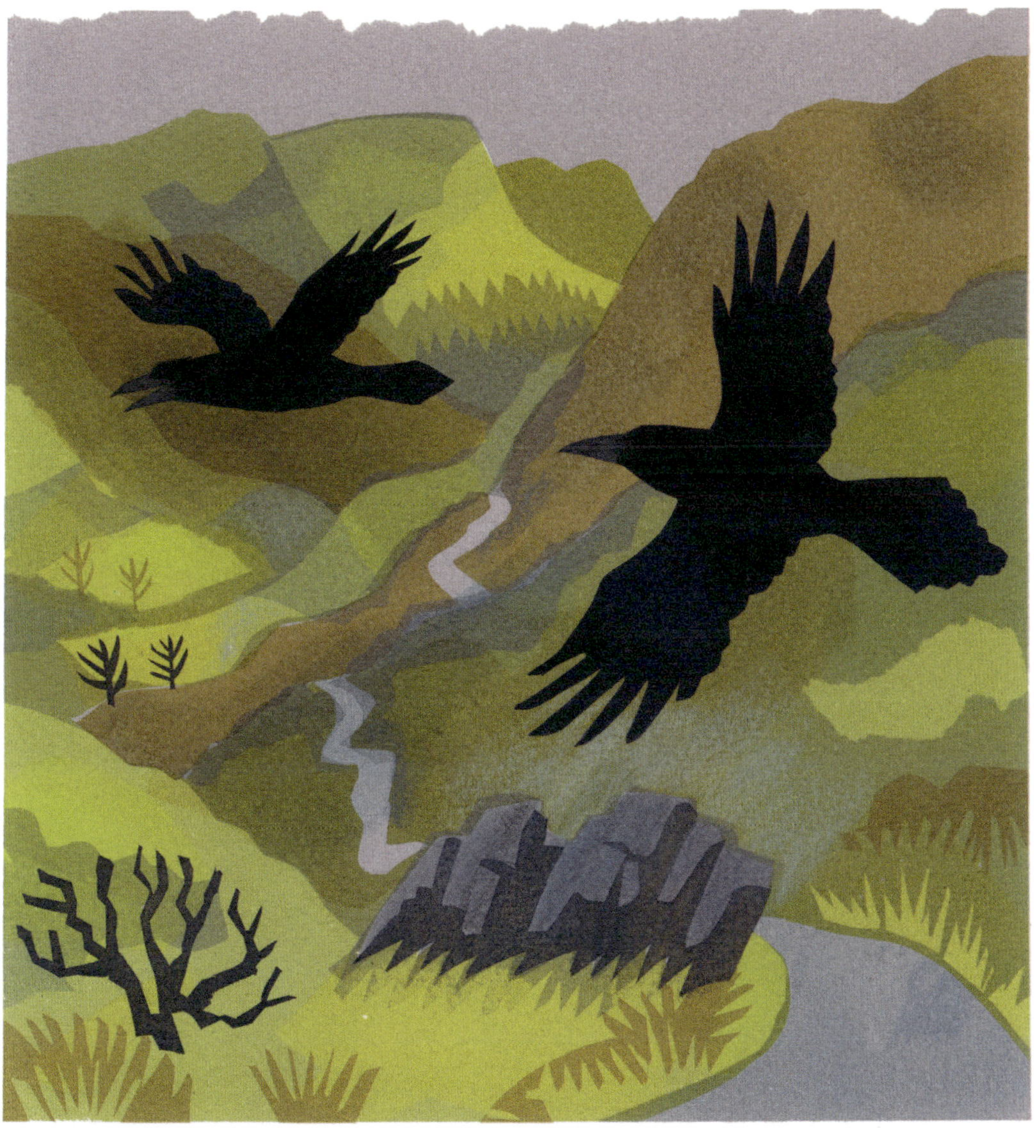

The **ravens** (*Corvus corax*) at the Tower of London are probably the world's most famous birds. Installed by Charles II, the belief that if they leave, Crown and country are doomed, is a Victorian invention. Nevertheless, Winston Churchill (1874–1965), during London's 1940 wartime bombing blitz, raised the garrison to full strength (six, plus a reserve). The ravens have military ID cards and are dismissible by the Ravenmaster (a post established in 1968). In 1986, Raven George was banished to the Welsh Mountain Zoo for destroying TV aerials. Raven Merlina, a favourite with the public, was a TV celebrity with a substantial social media following. The longest-lived Tower raven died at 44.

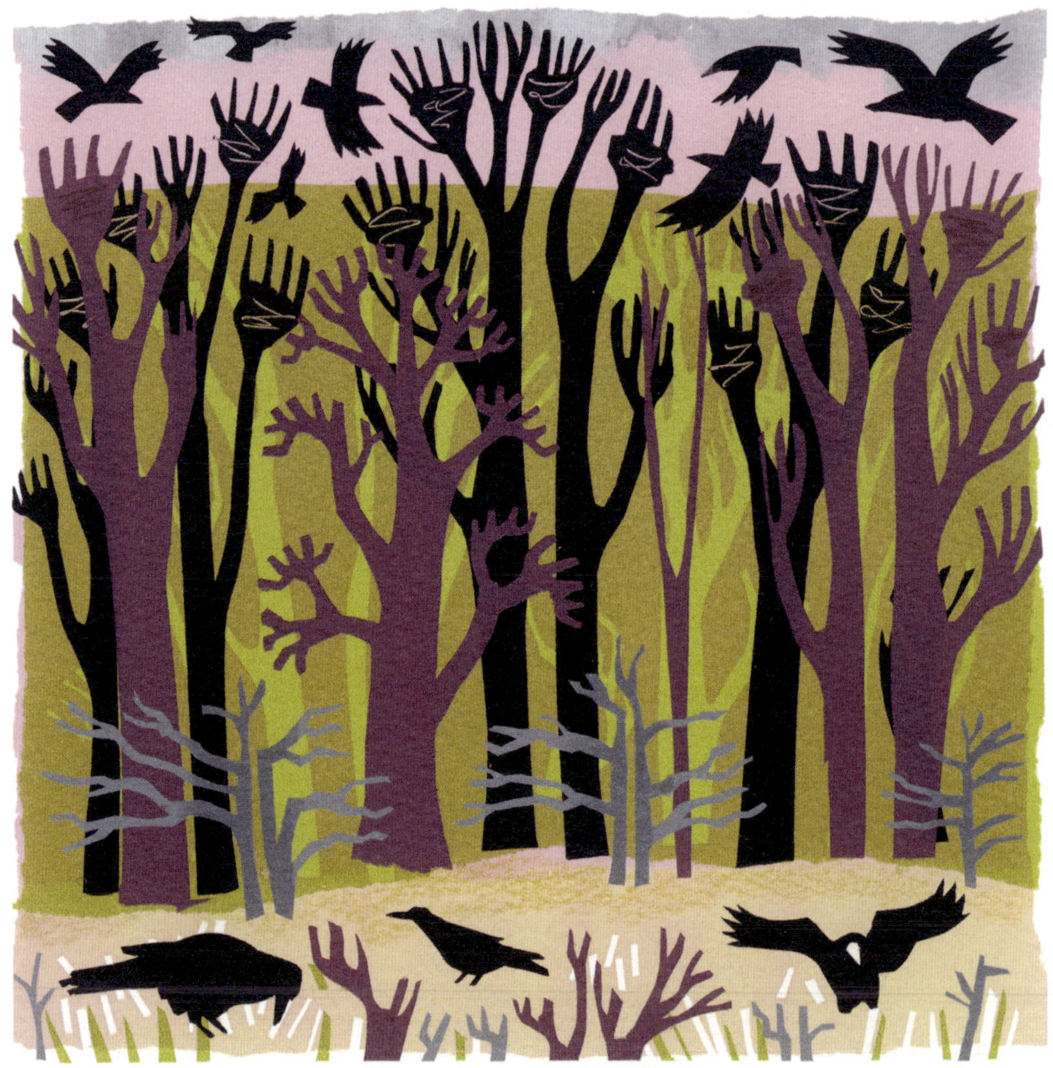

Carrion crows are solitary; the **rook** (*Corvus frugilegus*) is congregational. It has given its name to more British places than any other bird. Rookeries rarely comprise more than 100 nests and are often adjacent to villages. It is an ill omen if a rookery is abandoned of the birds' accord, although not, as in an un-named East Anglian rookery in 2022, because herring gulls ate all the chicks. Outside the nesting season, rooks roam in large flocks, often with jackdaws, to woodland roosts. They add to the drama of crop circles when their coal-black plumage catches the light as they feed on the flattened corn, turning it glinting black.

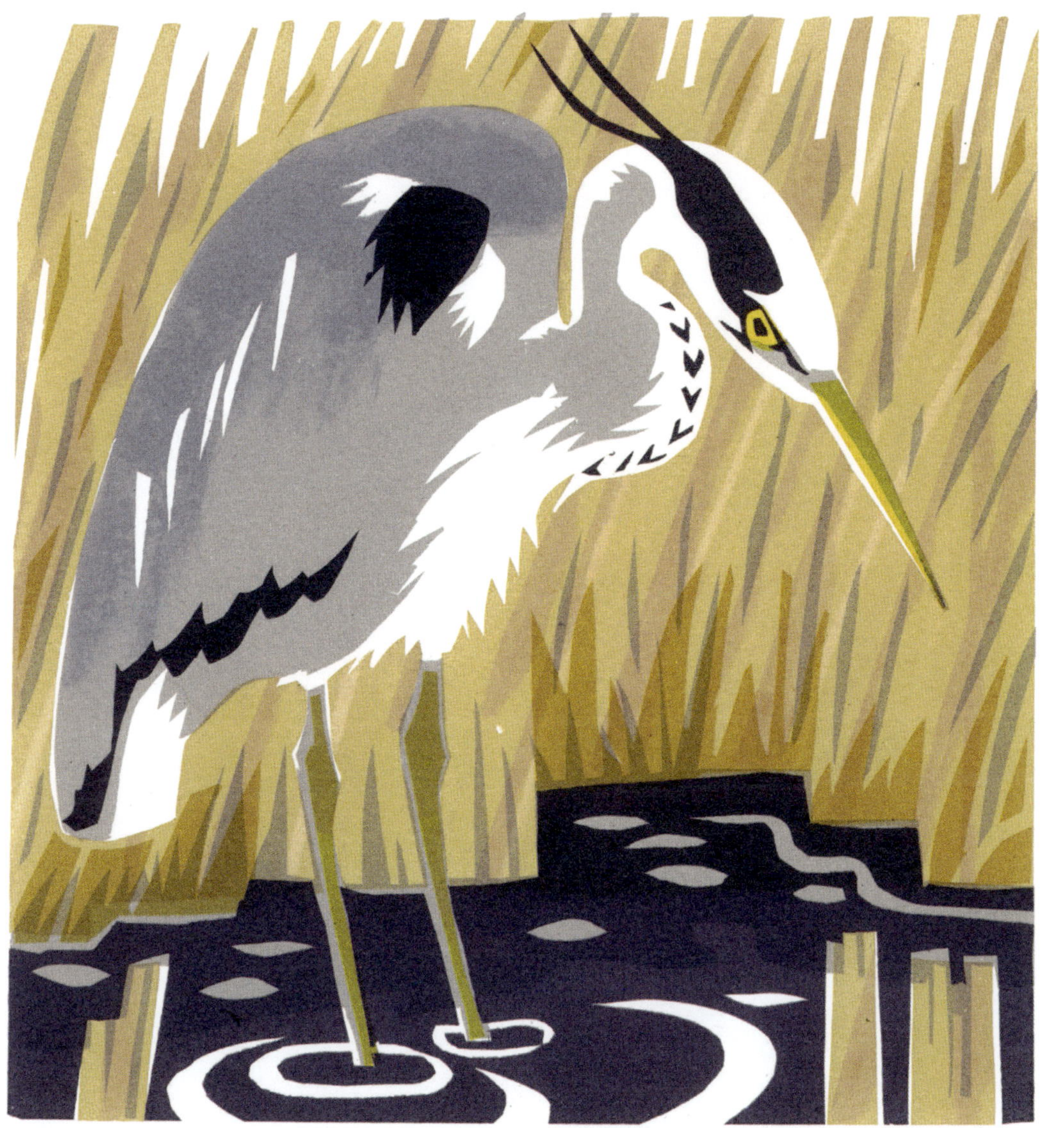

The **grey heron** (*Ardea cinerea*) gathers in a colony to breed, wedging its stick-piled nest in the branches of a tree. In treeless regions, it nests on cliffs, islets, even on moorland. In the British Isles, there are heronries from the Isles of Scilly to the Shetlands. The best known are on two islands in Regent's Park, London, the first nest appearing in 1968. Regent's Park herons scavenge the daily fish hand-outs to the Humboldt penguins at nearby London Zoo. Until they master the art of fishing, young herons practise on easier earth-bound or surface prey – rodents, frogs and chicks.

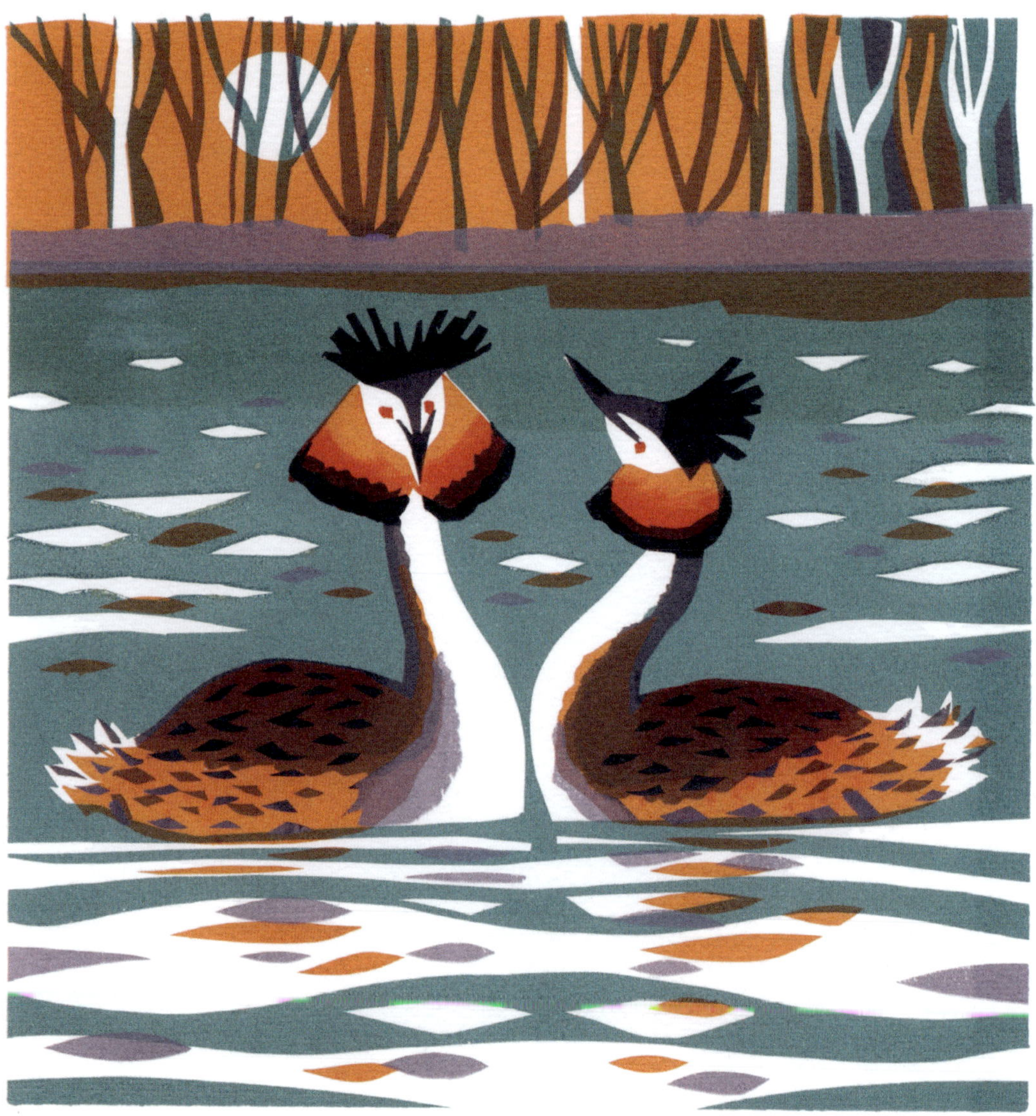

The spring courtship of the **great crested grebe** (*Podiceps cristatus*) demonstrates the similarity of human emotion to that of our fellow creatures. That was the conclusion of Sir Julian Huxley (1887–1975) in his landmark 1914 scientific paper, *The Courtship Habits of the Great Crested Grebe.* The birds end their aquatic ballet standing breast to breast, proffering beaks carrying wet weed – not dissimilar, Huxley felt, to the way a courting human couple communicates. A great crested grebe is the comic catalyst in *Scoop*, the satirical novel on journalism by Evelyn Waugh (1903–66). He resented the bird, having once been forced to go out after a long lunch to see its nest.

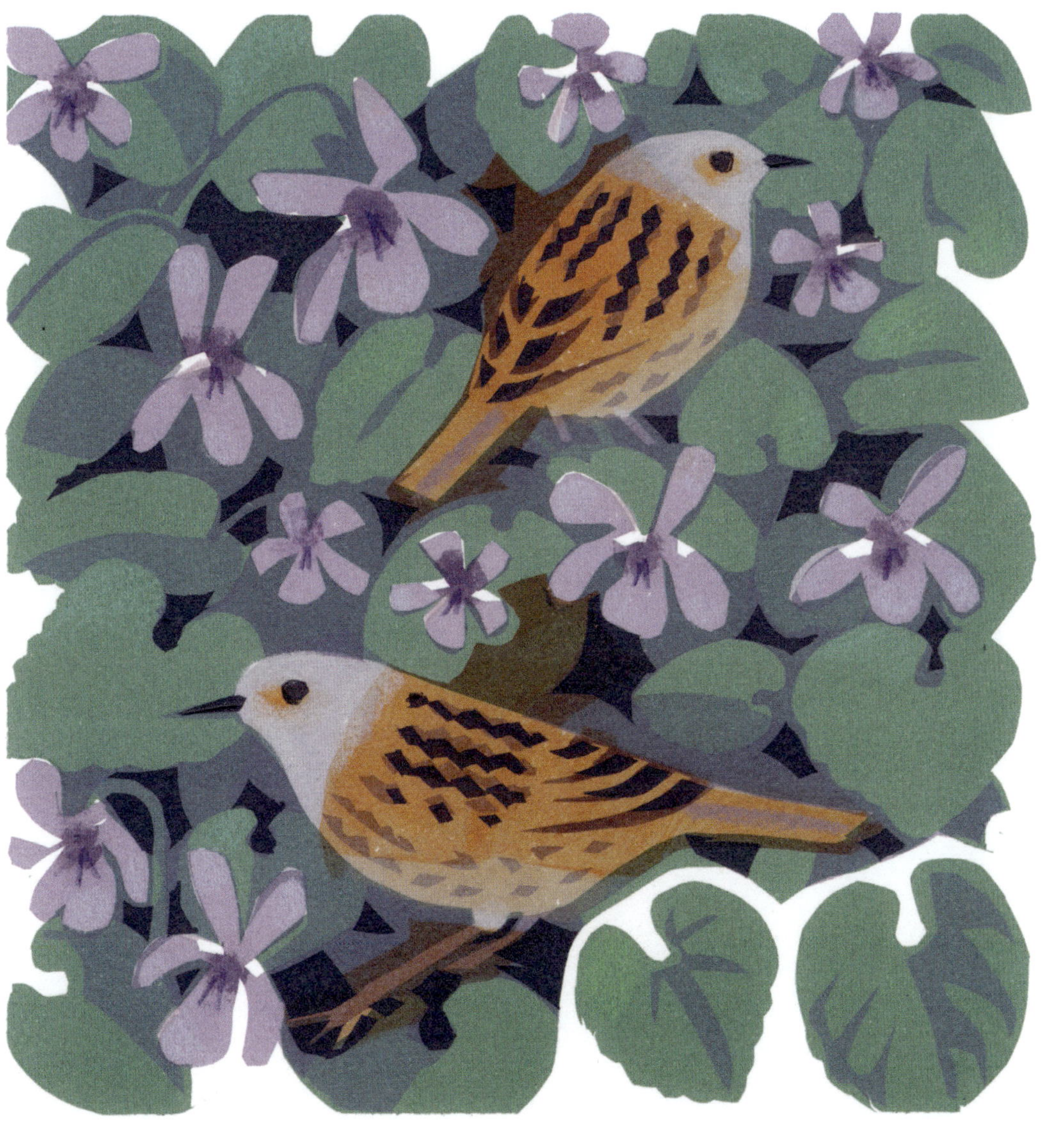

Charles Darwin (1809–82) knew female birds could have multiple breeding partners but wrote otherwise – not least because his daughter Etty corrected his page proofs. His naturalist contemporary, Rev. Francis Orpen Morris (1809–93), accepted Darwin's published views and particularly urged his congregation to follow the marital devotion of the **dunnock** (*Prunella modularis*). Subsequent research has made the female dunnock's promiscuity legendary. In 1951, ornithologist and eminent civil servant, Max Nicholson (1904–2003) argued for a change of seven bird names, one of them the restoration of the traditional 'dunnock' for 'hedge sparrow'. Dunnock was the only acceptance. It is not a sparrow but an accentor (*Prunella*).

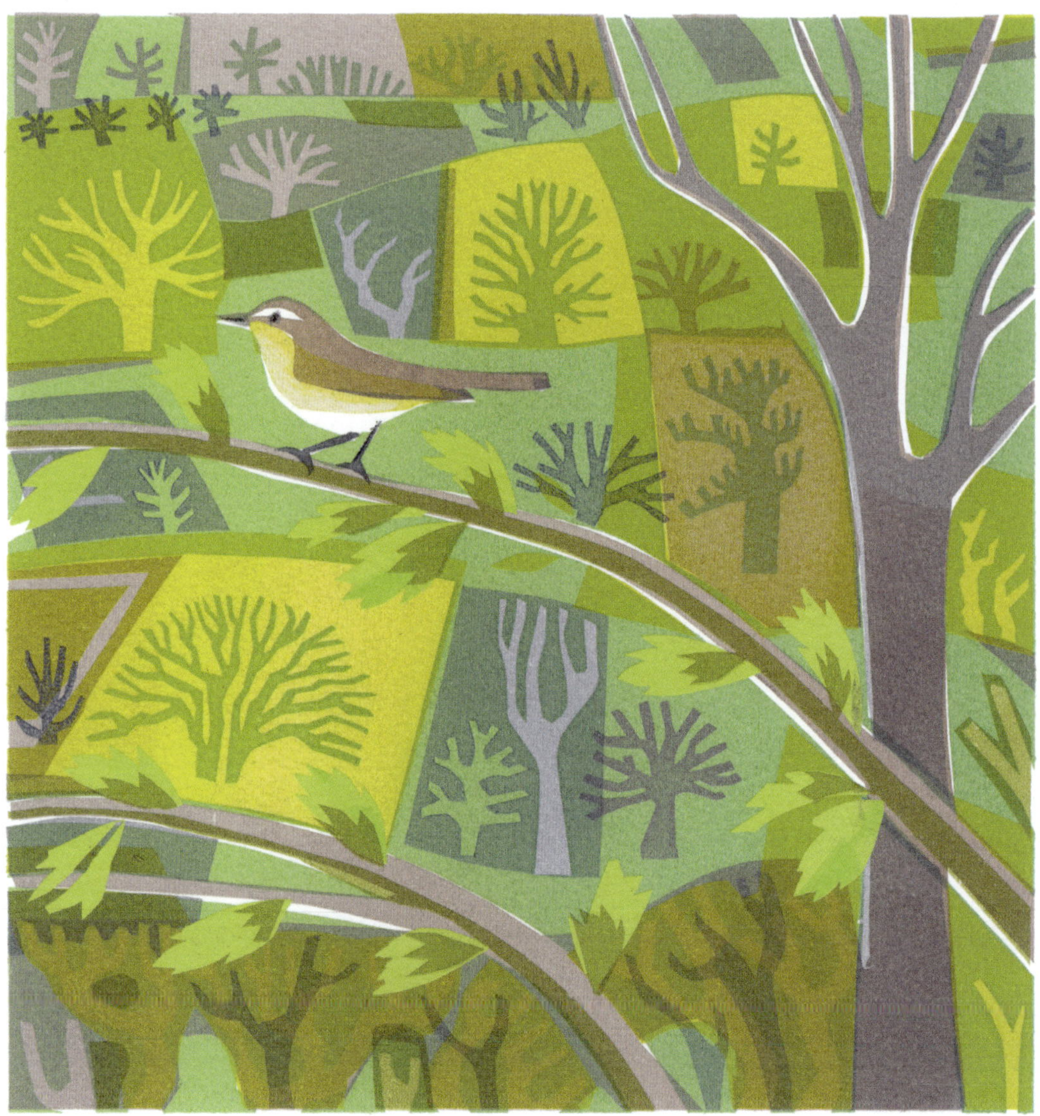

Of the millions of summer migrant birds to the British Isles from Africa and southern Europe, the most identifiable is the **chiffchaff** (***Phylloscopus collybita***), named after its repeated call. The ornithologist H. F. Witherby (1873–1943) transcribed its song: 'chiff-chaff-chaff-chaff-chaff-chaff-chiff-chiff-chiff-chaff-chaff-chiff-chiff-chiff-chaff' (*The Handbook of British Birds*, Vol II). Onomatopoeic attempts in other languages include Welsh *siff-saff*, Dutch *tjitjat* and German *zilpzalp*. It is a 'leaf' warbler, and its distinctive call announces its presence. A visible difference between the chiffchaff and the very similar willow warbler is that the former has black legs. As a reminder, there is a saying: 'Chiffchaff riffraff with dirty legs'. It is the first warbler to arrive in spring, boosting a growing resident UK population of 1,000 birds to more than a million.

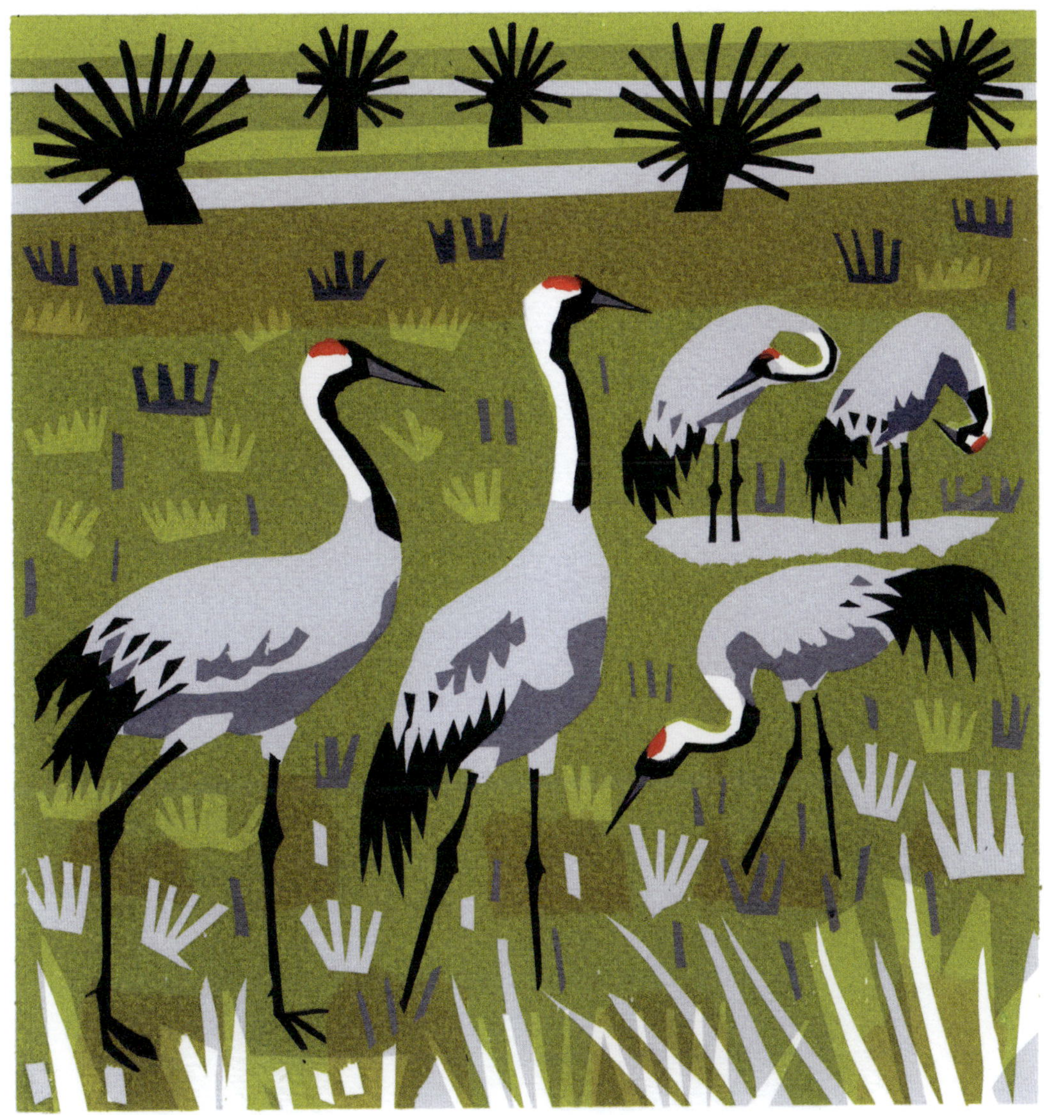

Cranberries, cranesbill, and hundreds of British place names from Cranbrook (Kent) to Cran Hill (Aberdeenshire) testify to the once widespread population of the **crane** (*Grus grus*), the largest British bird at up to 120cm (4ft) tall with a 245cm (8ft) wingspan. Yet when a chick fledged at Horsey, Norfolk Broads, in 1982, it was the first in the British Isles for 400 years. Its parents had wandered off the western migration from Spain and southern France to Scandinavia. The owner of Horsey, John Buxton (1927–2014), wildlife film-maker and confessed craniac, successfully protected what became a small resident colony of the shy and wary birds from egg stealers, foxes and uninvited birdwatchers (John Buxton & Chris Durdin, *The Norfolk Cranes' Story*). Horsey inspired the Great Crane Project, an imported introduction focused on the Somerset Levels. In 2024 there are 145 wild UK residents.

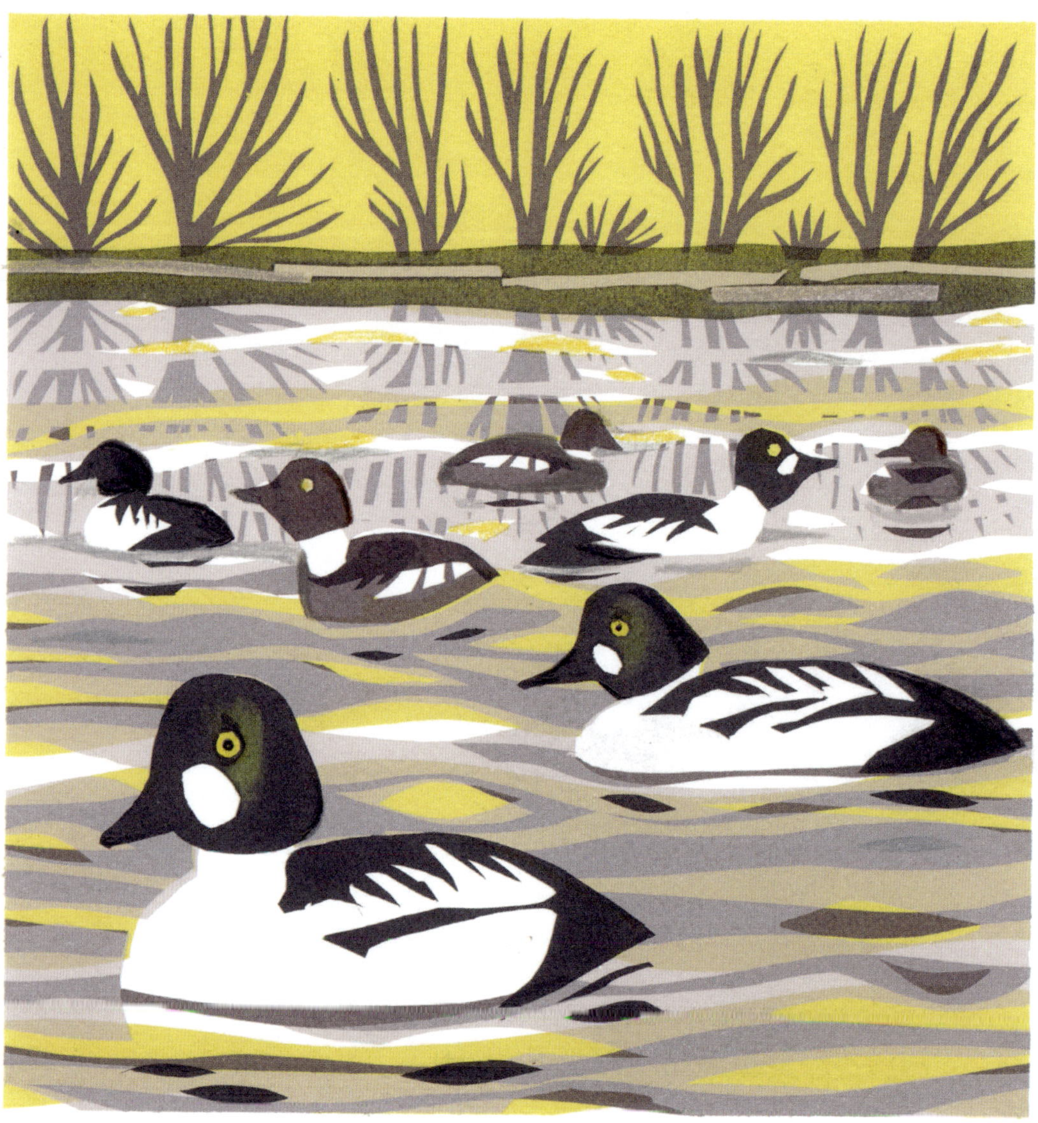

The three diving ducks most commonly seen in inland parts of the UK have populations that are greatly increased by winter migrants: tufted (38,000–140,000), pochard (1,440–29,000) and **goldeneye** (*Bucephala clangula*) (400–21,000). The goldeneye nests in tree holes. Laplanders have used nest boxes for 200 years to harvest their eggs more efficiently. Since 1970, nest boxes have been provided in the Spey Valley in the Highlands to encourage goldeneyes to breed, to good effect – 200 pairs now breed in the area and beyond. Before the goldeneyes' March exodus, drakes can begin courting – laying their heads back, kicking up fountains and uttering a piercing double whistle.

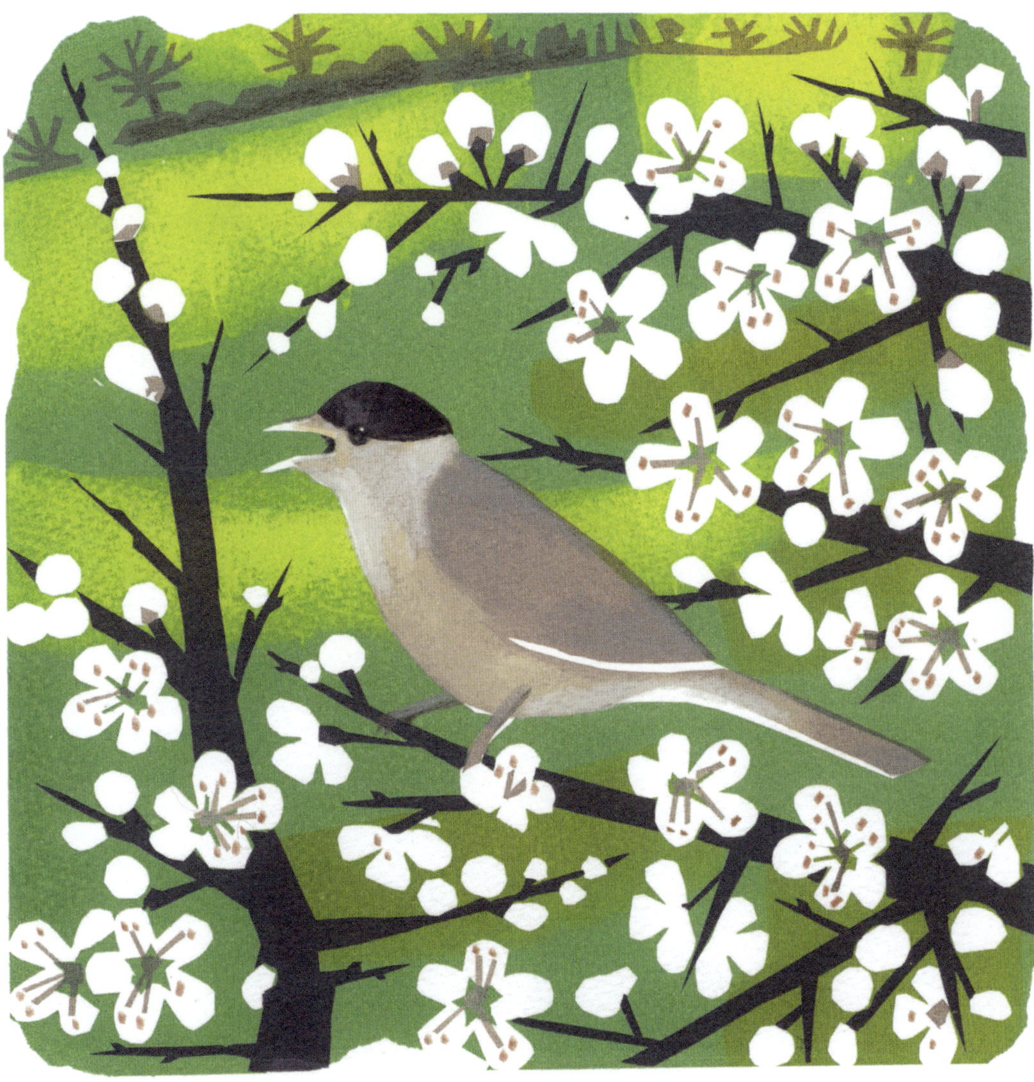

Most warblers have similar plumage. An exception is the **blackcap** (*Sylvia atricapilla*) with its distinctive 'skullcap', black for males, rust for females. This migrant has 1.65m breeding territories in the UK, replaced by 3,000 visitors from (mainly) central Europe in winter. Our wintering blackcaps are regular visitors to rural and urban bird tables. The male can be a 24-hour songster, though less markedly nocturnal than the nightingale. Gilbert White described its 'wild sweetness' but also noted a 'hurrying manner not at all to its advantage' in *The Natural History of Selborne*. Richard Smyth describes its call in *A Sweet, Wild Note* as 'Cracked ... littered with chitters and whistles, and generally all over the shop'.

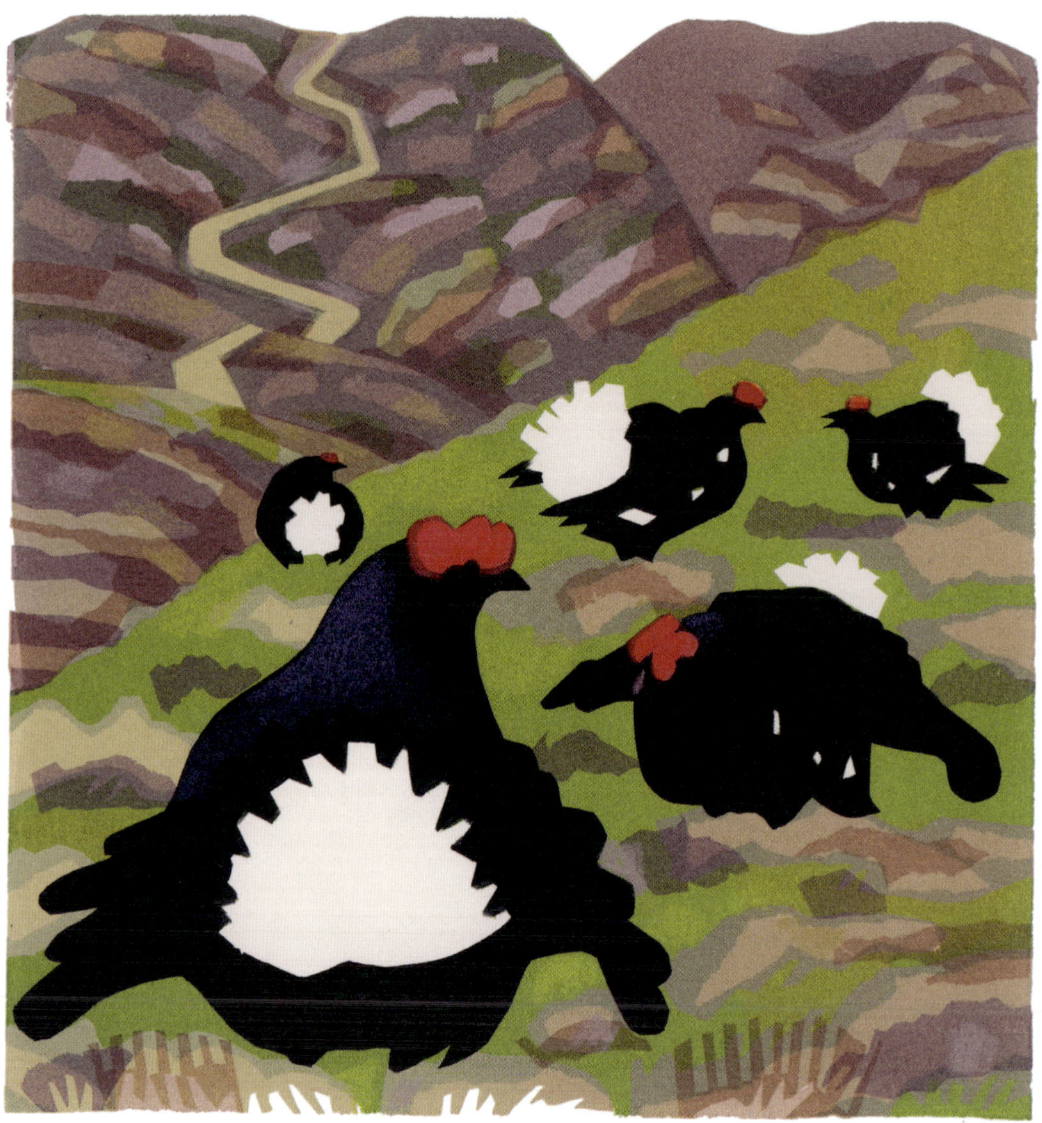

The **black grouse** (*Lyrurus tetrix*) is a contender for mainland Britain's most exotic bird, especially the male (or blackcock) with his lyre-shaped tail. The blackcock is at his most splendid showing off at a lek (from the Swedish *leka*, meaning 'to play'). In spring, blackcocks gather at dawn, tail-fanning and fighting to impress an audience of female greyhens. If a first-year cock has the effrontery to join in, he is chased off by his elders and betters. The most domineering blackcock has his way with more than one greyhen. Leks continue outside the mating season; the Angus lek depicted above was in September, hence no greyhens and postures rather than fights.

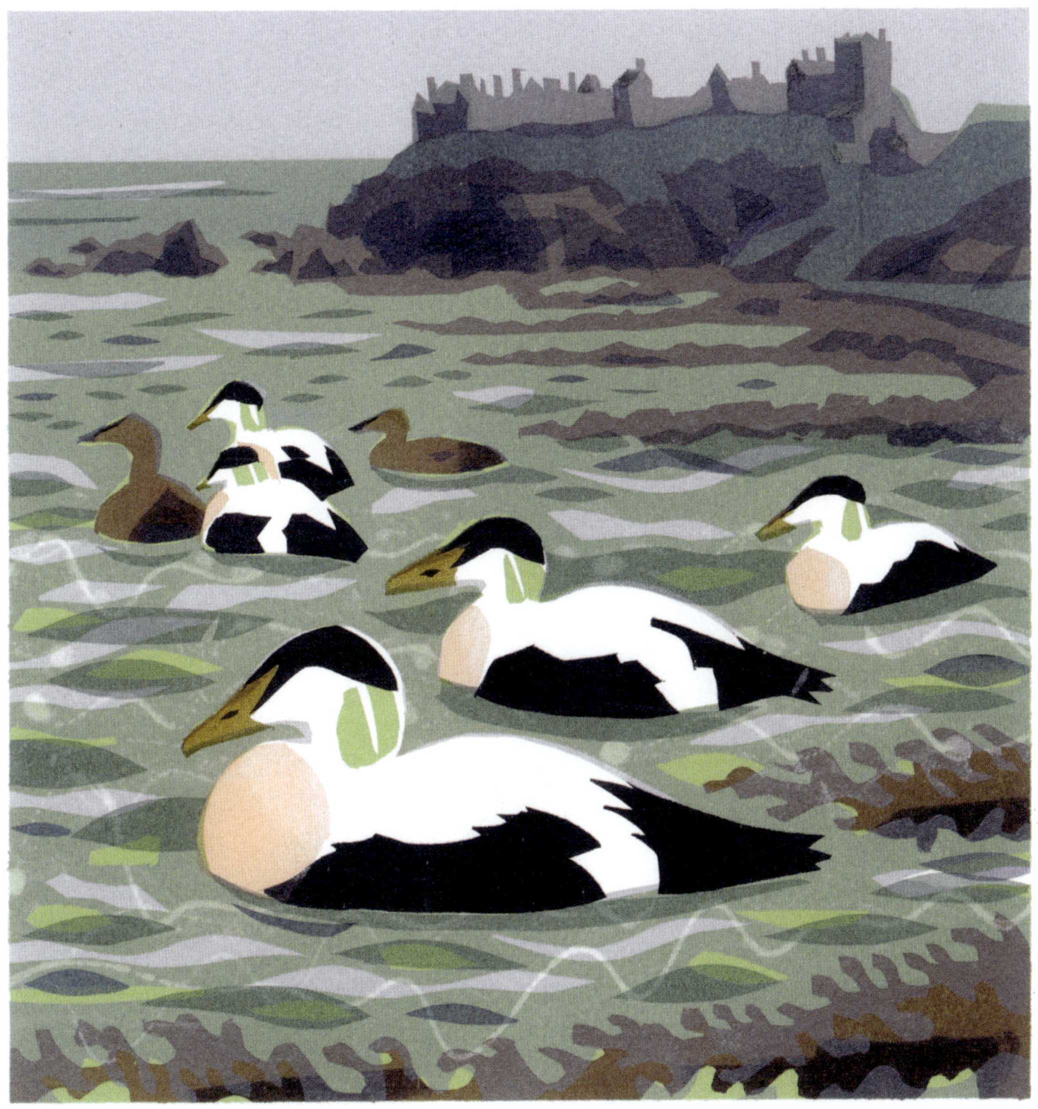

A population of 37,000 pairs makes the sea-bound **eider** (*Somateria mollissima*) the UK's commonest breeding duck, after the mallard, and its speediest. Coquet Island and the Farne Islands, off the Northumberland coast near Bamburgh Castle, mark the species' southern breeding limit. St Cuthbert (*c.* 634–687) introduced legal protection for all birds on the islands in what may be the first example of bird conservation. In Northumberland dialect, the eider is a 'Cuddy' (Cuthbert) duck. Eiderdown feathers remain the best insulation, withstanding cold to -35°C. Protected for their down, eiders are one of Iceland's most profitable resources. A single duvet filled with eiderdown can cost £4,000.

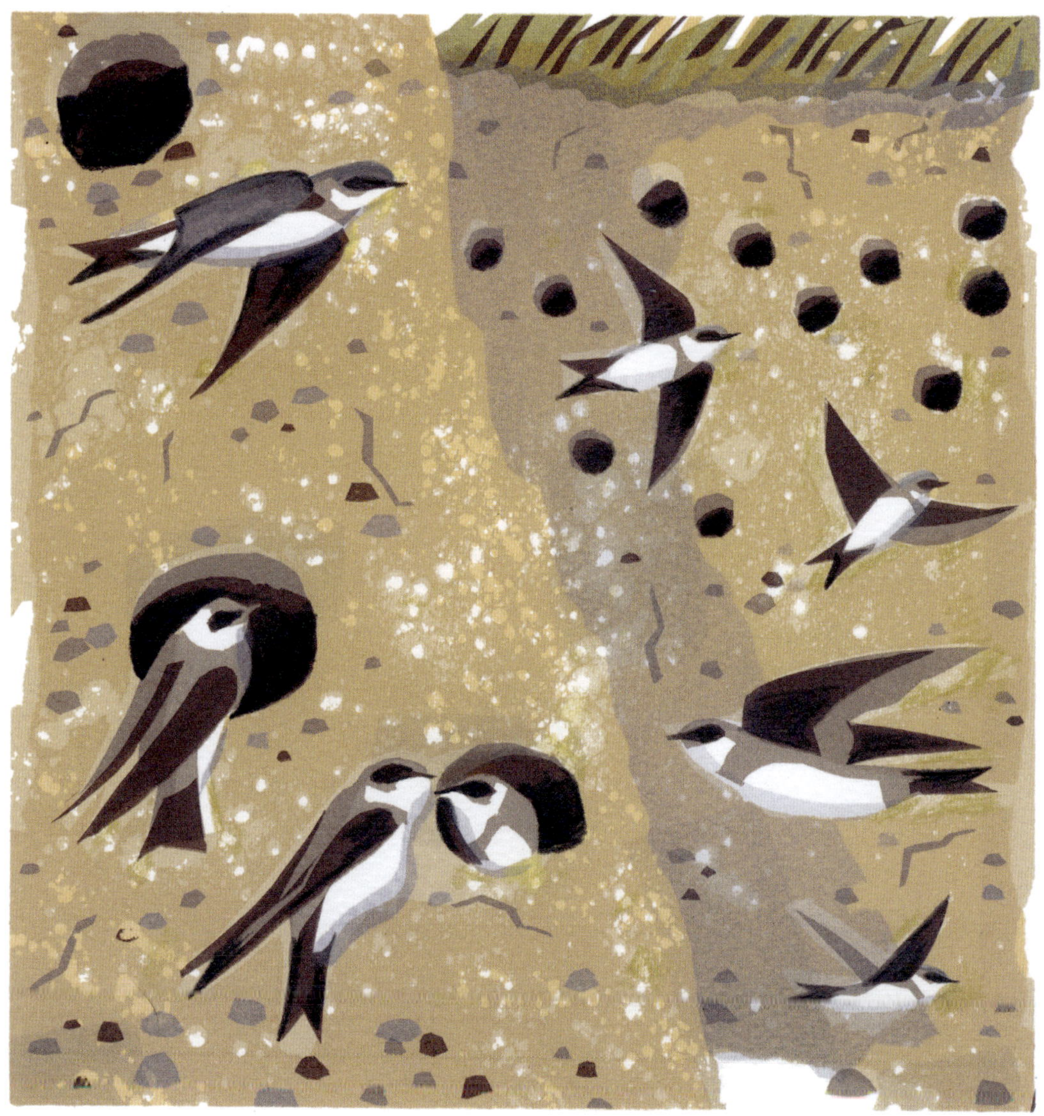

The **sand martin** (*Riparia riparia*) is the smallest, least numerous, but most gregarious of the swallow/martin family (Hirundinidae), honeycombing sand banks inland and coastal with tunnelled nests, as the poet John Clare (1793–1864) noted in 'The Sand Martin':

More like the haunts of vermin than a bird

It will hunt over urban waters but is the least human-friendly of its family:

Ive seen thee far away from all thy tribe
Flirting about the unfrequented sky
And felt a feeling that I cant describe
Of lone seclusion and a hermit joy

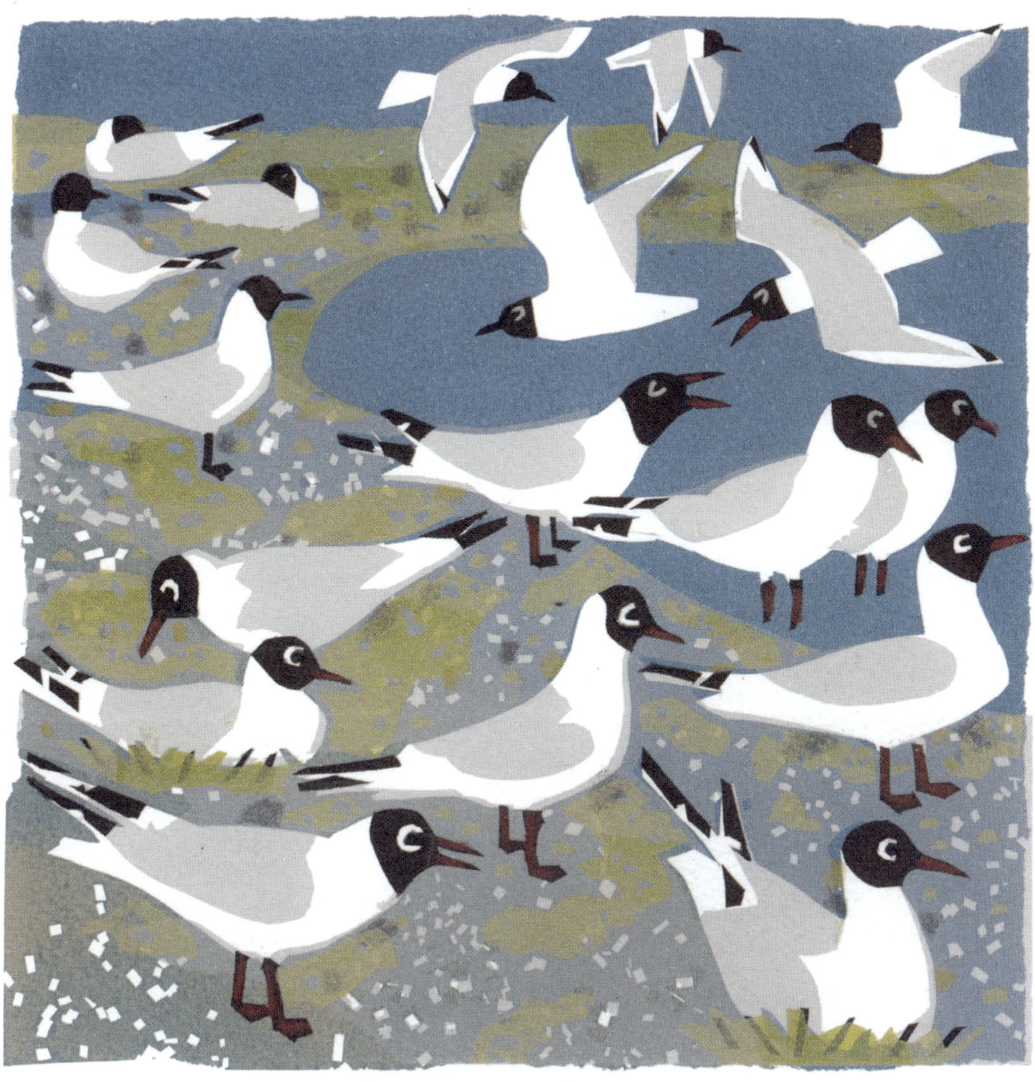

Epicures value the **black-headed gull** (*Chroicocephalus ridibundus*) for its eggs, the only ones of a wild bird licensed for harvest (quail and pheasant eggs are farmed). In winter, with 2.2m present, black-headeds are three times more numerous than any other British gull (the UK also has 130,000 resident pairs). Early and aggressive nesters – dark-brown heads reappearing in February – they deprive summer-migrant terns of nest sites and, like other gulls, will prey on other birds' chicks. In winter children find them a delight, when residents return from wetland breeding colonies to join the migrant influx on urban waters. Tossed scraps of bread (preferably wholegrain) caught on the wing reveal their aerial agility.

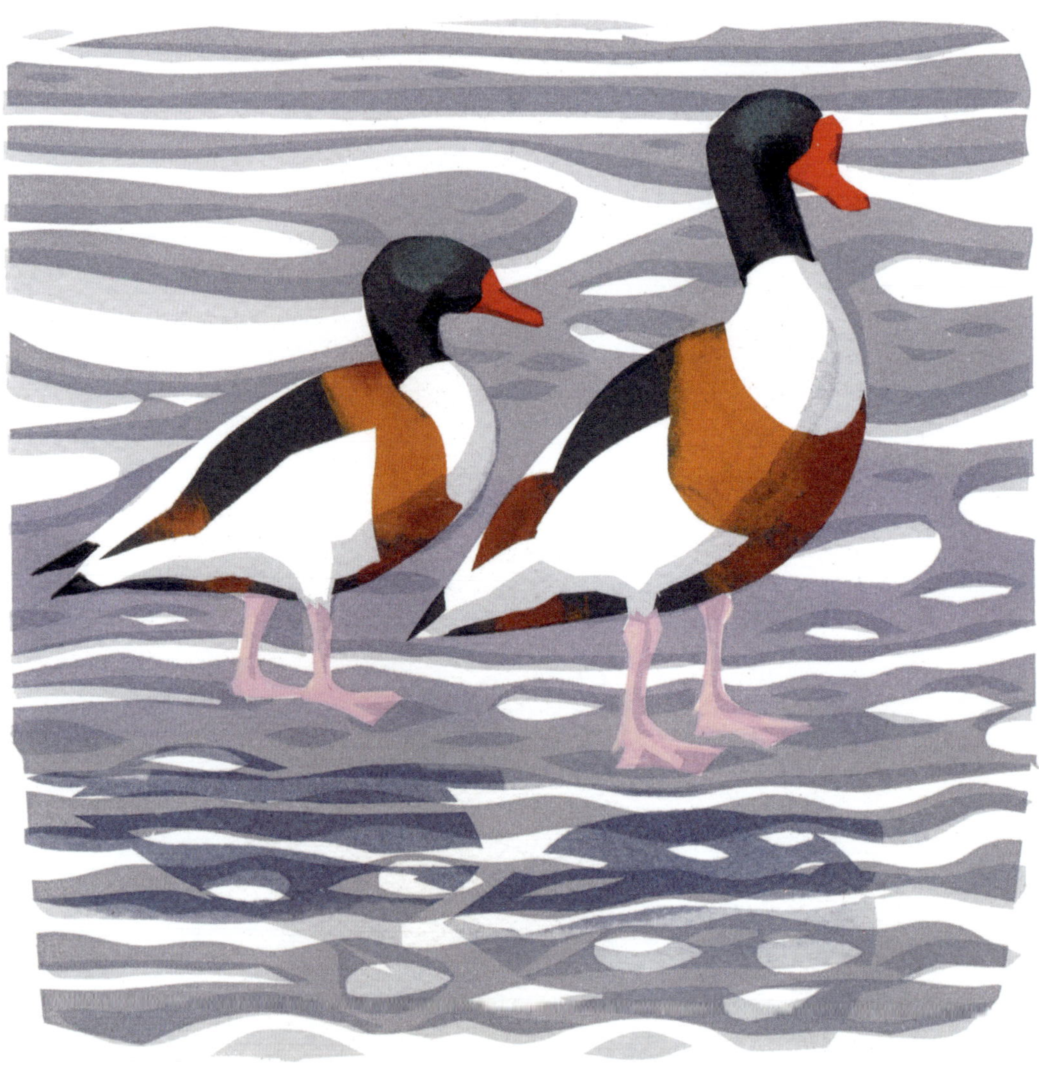

The widening winter range of the **shelduck** (*Tadorna tadorna*) to include most of Britain's coastline, brightens many a bleak journey with its striking plumage ('sheld' comes from the Middle Dutch *schillede*, 'variegated'). The eye-catching effect applies to both sexes – camouflage is unnecessary as it nests in burrows. Unusual among ducks are the drake's territorialism, that burrowless females add their eggs to existing nests (32 the record clutch), and 'creches' of up to 100 ducklings supervised by a single adult pair. The most goose-like duck, local names include 'skeel goose', 'links goose' and 'sly goose'.

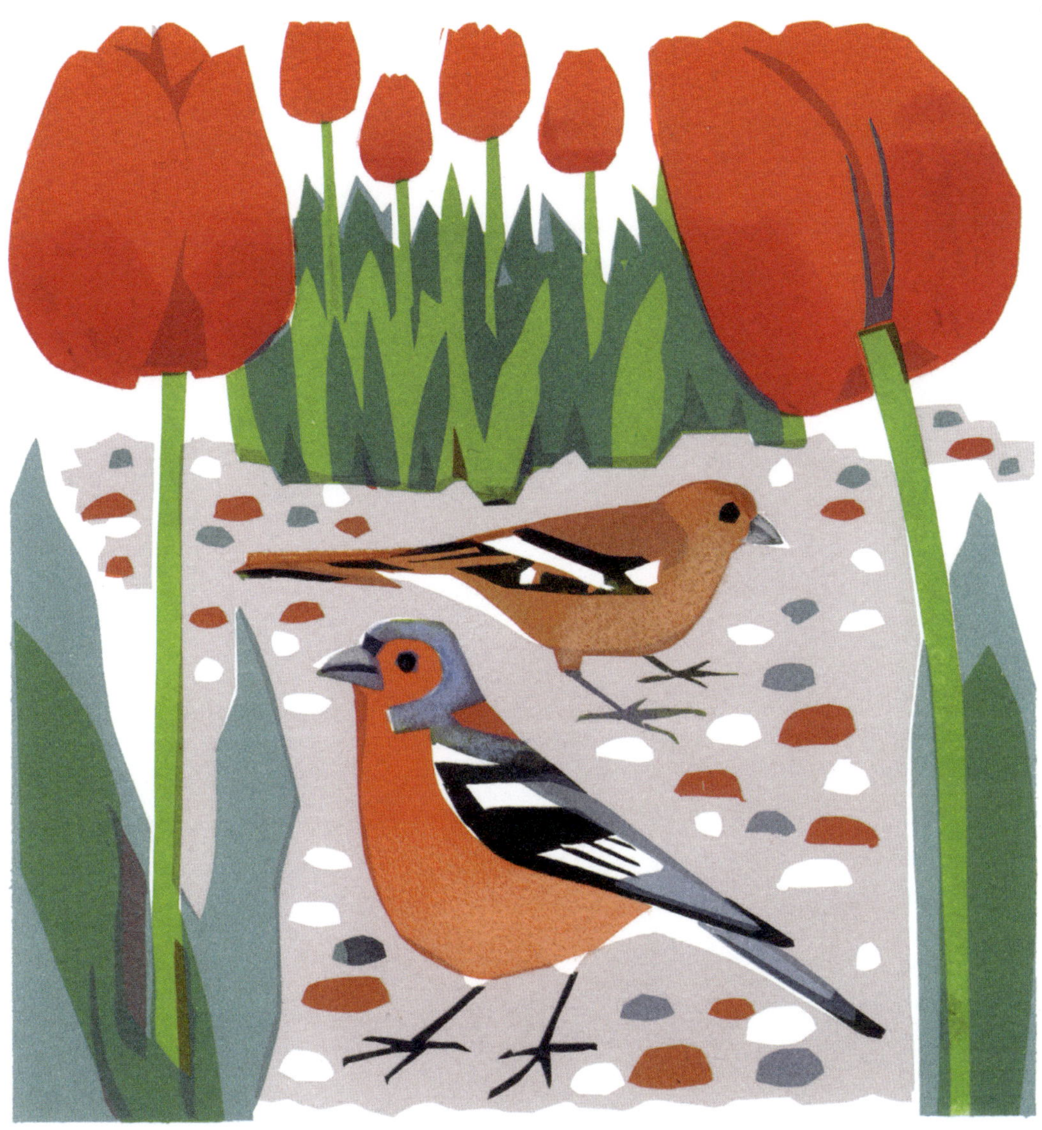

The **chaffinch** (*Fringilla coelebs*) is joint fifth with the blackbird in the list of 22 bird species with UK populations over a million. In stubble-field days, more numerous. Henry Douglas-Home cursed them for interrupting bird recordings. The moss- and lichen-clad nest with its feathered interior is the female's masterpiece, the resplendent male chief gatherer.

Chaffinch carries the moss in his mouth
To filbert hedges all day long
And charms the poet with his beautiful song

John Clare, from 'Birds Nests' (his last poem)

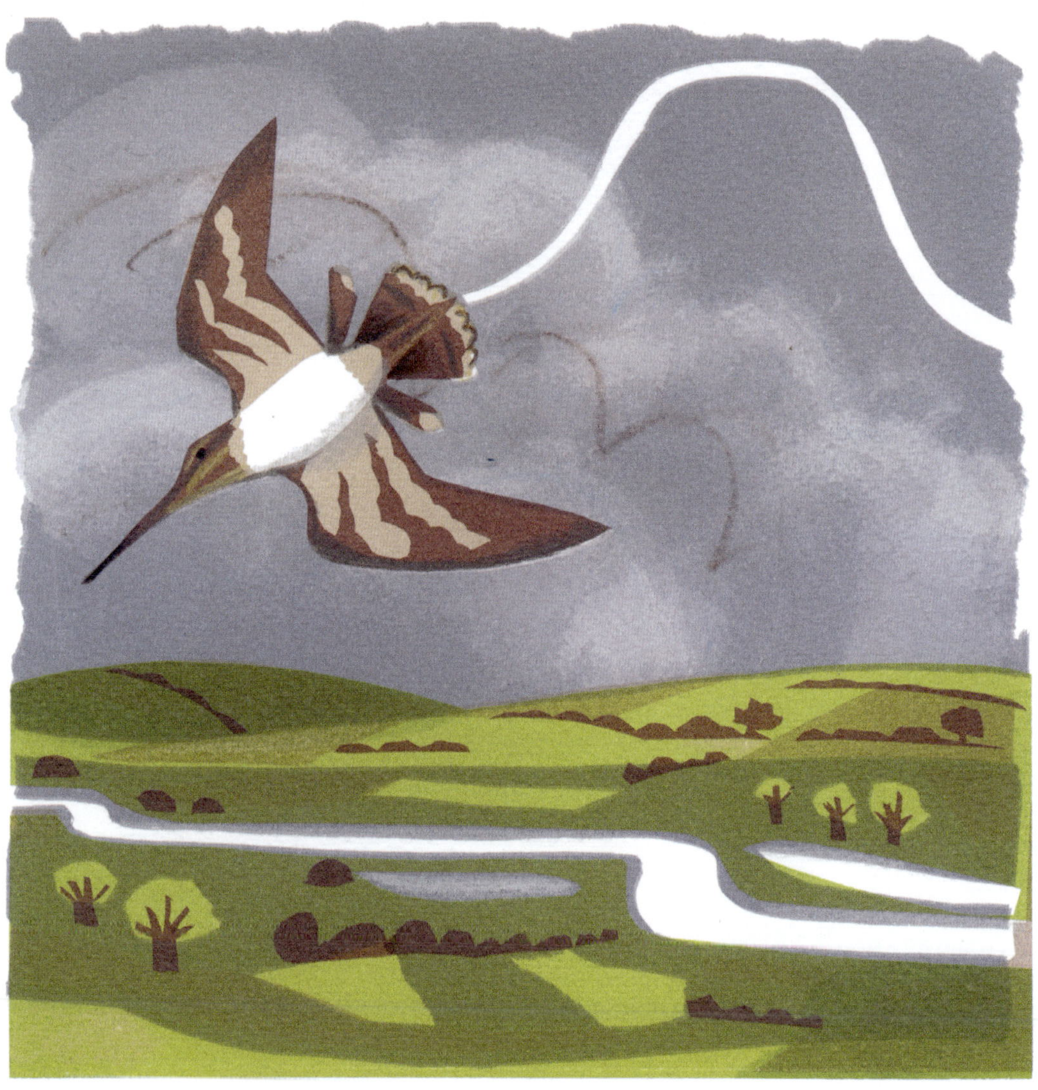

The **snipe** (*Gallinago gallinago*), an inland wader of bogs and marshes, has proportionately the longest bill, at one-third of its body length, of any British bird. In spring, males stake their territory with a 'drumming' flight: a bleating noise caused by the outer tail feathers vibrating during dives. A flushed snipe utters a rasping call and, despite its zig-zag flight, reaches 100km/h (60mph) in seconds. Bird hides can afford scrutiny of the snipe's grassy camouflage, hence the word 'sniper' for a hidden marksman. Lawyers earned the nickname 'snipe' for their lengthy bills.

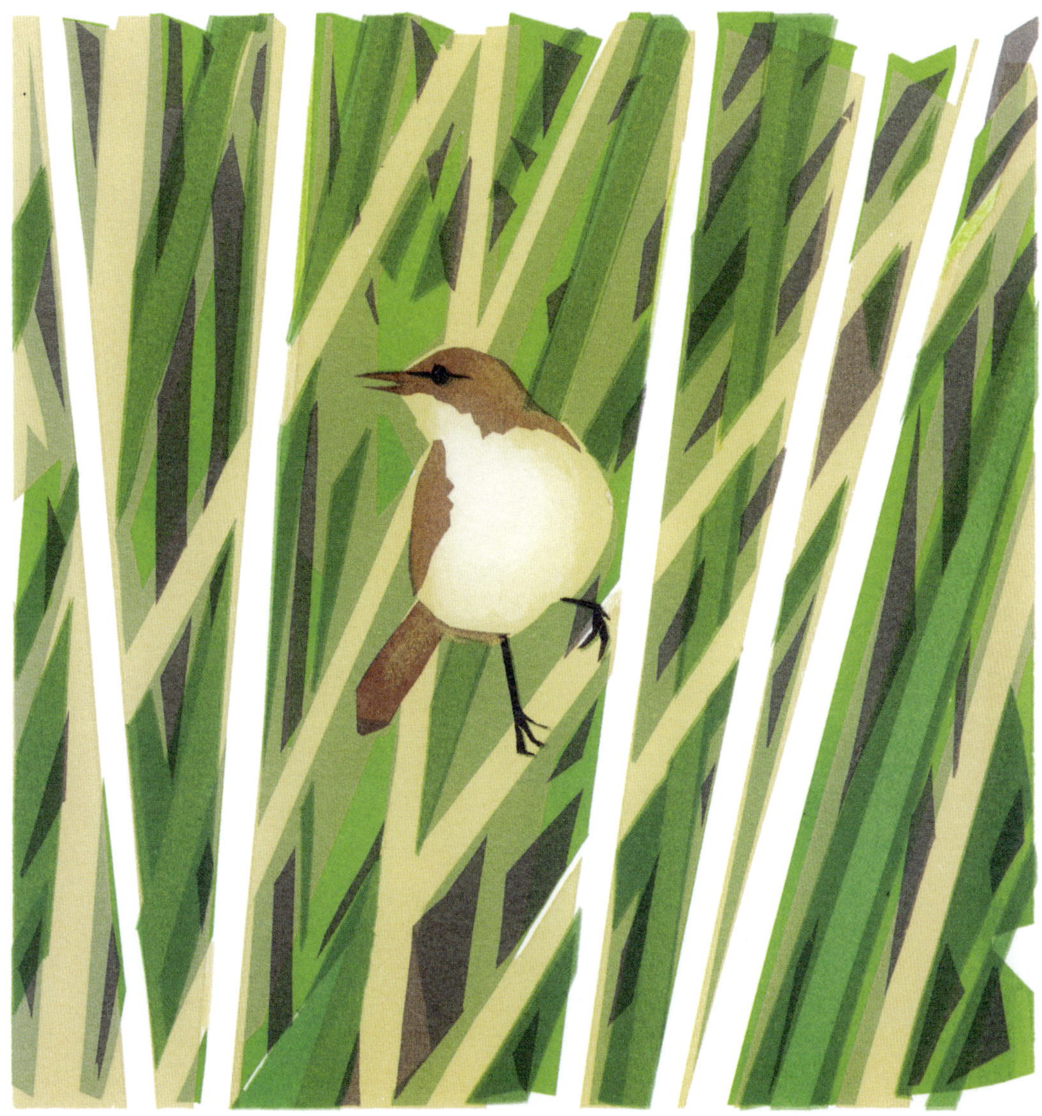

Reed beds bind loose soil, absorb pollutants, prevent water erosion and enable biodiversity, but 90 per cent of UK reedbeds have disappeared over the last 100 years. Legal steps for reedbed reintroduction have been taken globally since 1992. One beneficiary is the summer migrant **reed warbler** (*Acrocephalus scirpaceus*), a favourite host of the cuckoo. UK numbers have since doubled to 130,000 pairs, and breeding populations have appeared in London's Royal Parks, as well as Ireland and Scotland. Its trills and scratchy warblings are even audible within yards of oblivious carousers and bathers at London's Serpentine Lido in Hyde Park.

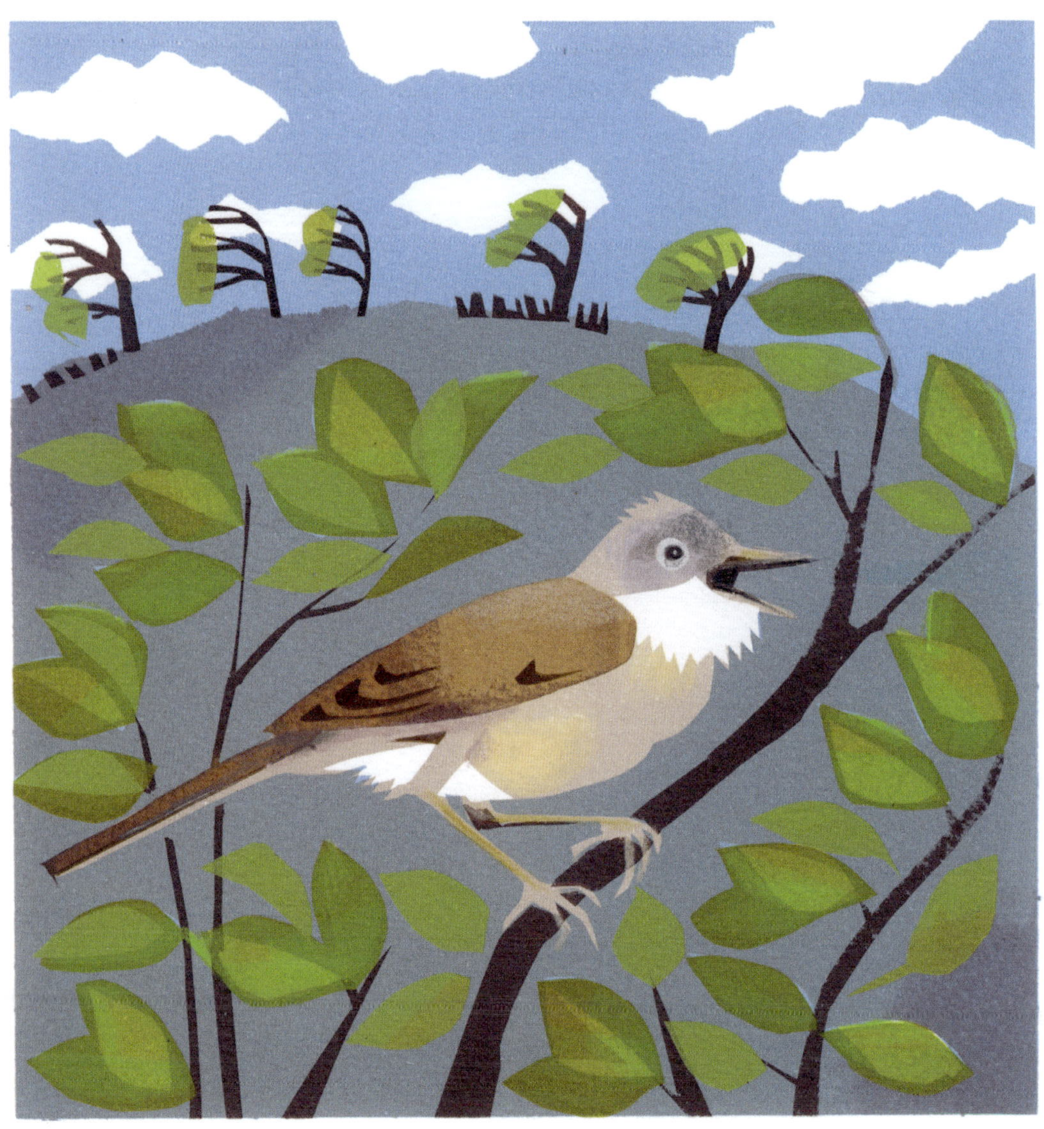

In 'The Happy Bird', John Clare rightly identifies the summer migrant **whitethroat** (*Sylvia communis*), with 1.1m hedgerow and scrub territories, by its song:

The happy white throat on the sweeing bough
Swayed by the impulse of the gadding wind
That ushers in the showers of april – now
Singeth right joyously and now reclined
Croucheth and clingeth to her moving seat
To keep her hold – and till the wind for rest
Pauses – she mutters inward melodys
That seem her hearts rich thinking to repeat

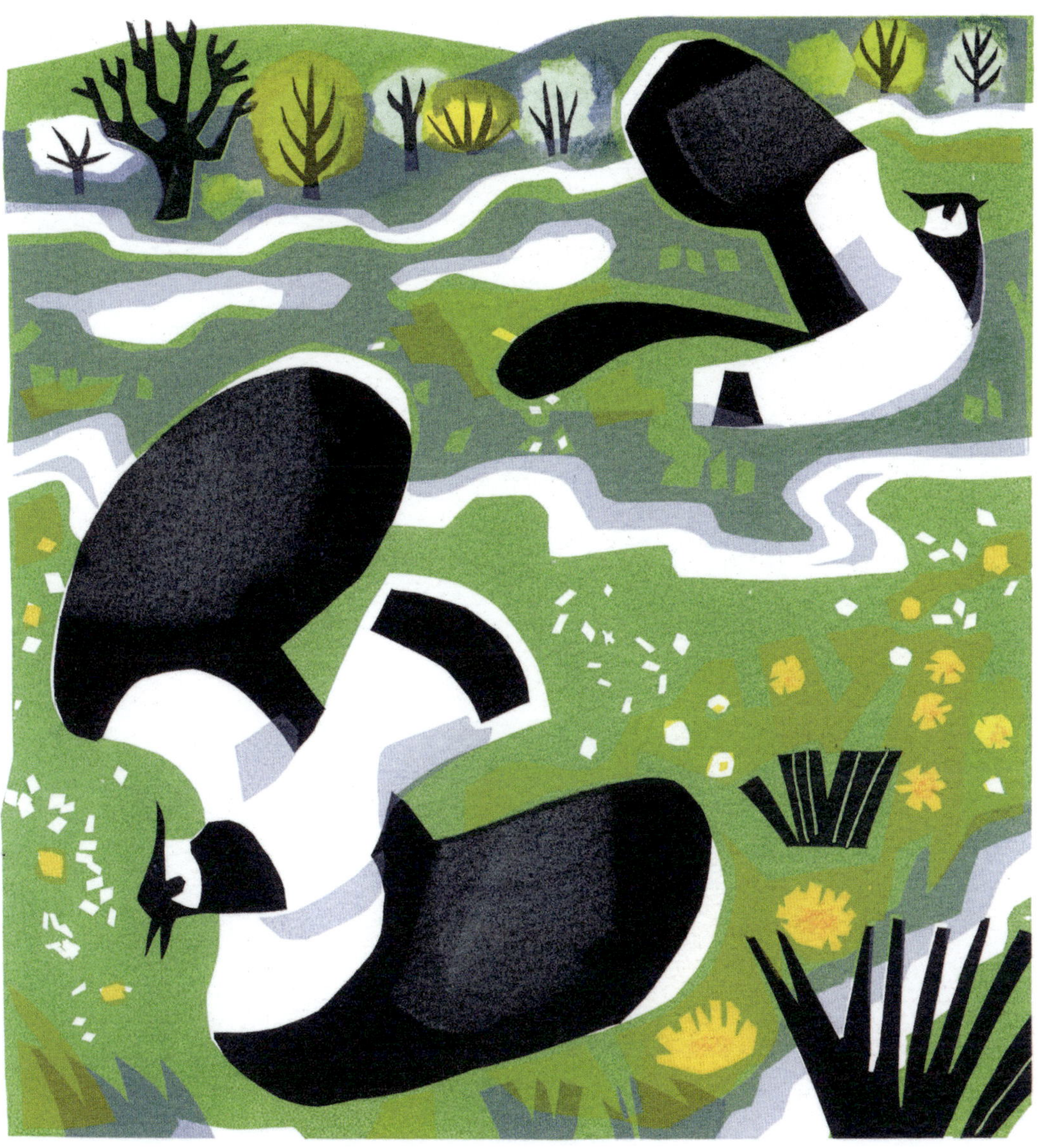

The wing of the **lapwing** (*Vanellus vanellus*, meaning 'little fan'), used by the male to aerobatic effect in the spring, helped inspire the design of the Spitfire fighter plane by R. J. Mitchell (1895–1937). Its alternative name, 'peewit', and other regional variants imitate its call. From faraway Samoa at the end of his life, Robert Louis Stevenson remembered them:

Be it granted me to behold you again in dying,
Hills of home! And to hear again the call;
Hear about the graves of the martyrs the peewees crying;
And hear no more at all.

From 'To S. R. Crocket'

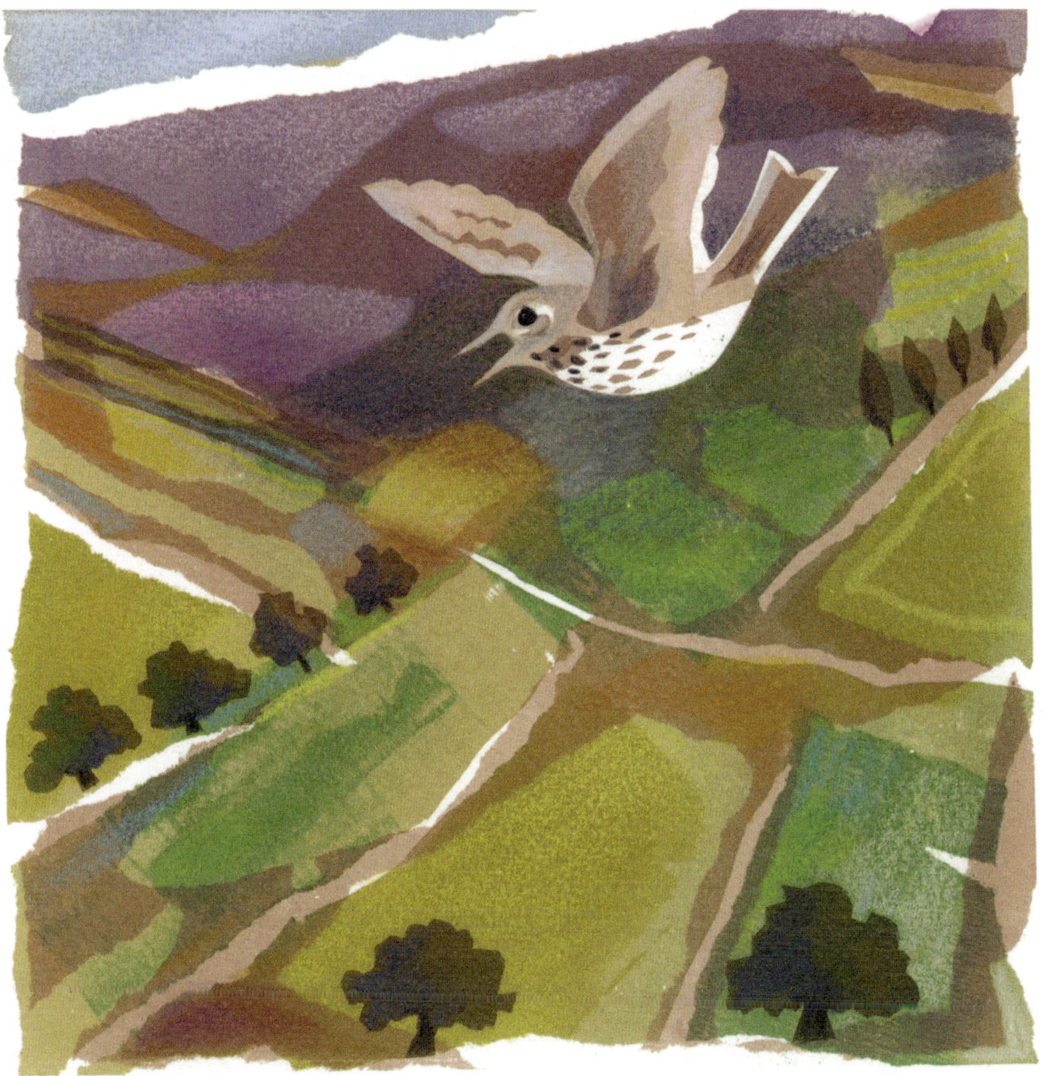

There are at least 35 different pipits, three of which are readily seen in the British Isles. **meadow pipit** (*Anthus pratensis*, at 2.75m UK pairs), rock pipit (36,000 pairs) and summer migrant tree pipit (145,000 pairs). Their similar appearance makes habitat the surest identification. This was demonstrated by the morale-boosting wartime comedy film *Tawny Pipit* in which the rare migrant was played by a meadow pipit. A westerly and northern breeding bird, Scottish and Welsh meadow pipits favour England in winter. Earlier southern springs make it a favourite foster parent for the cuckoo.

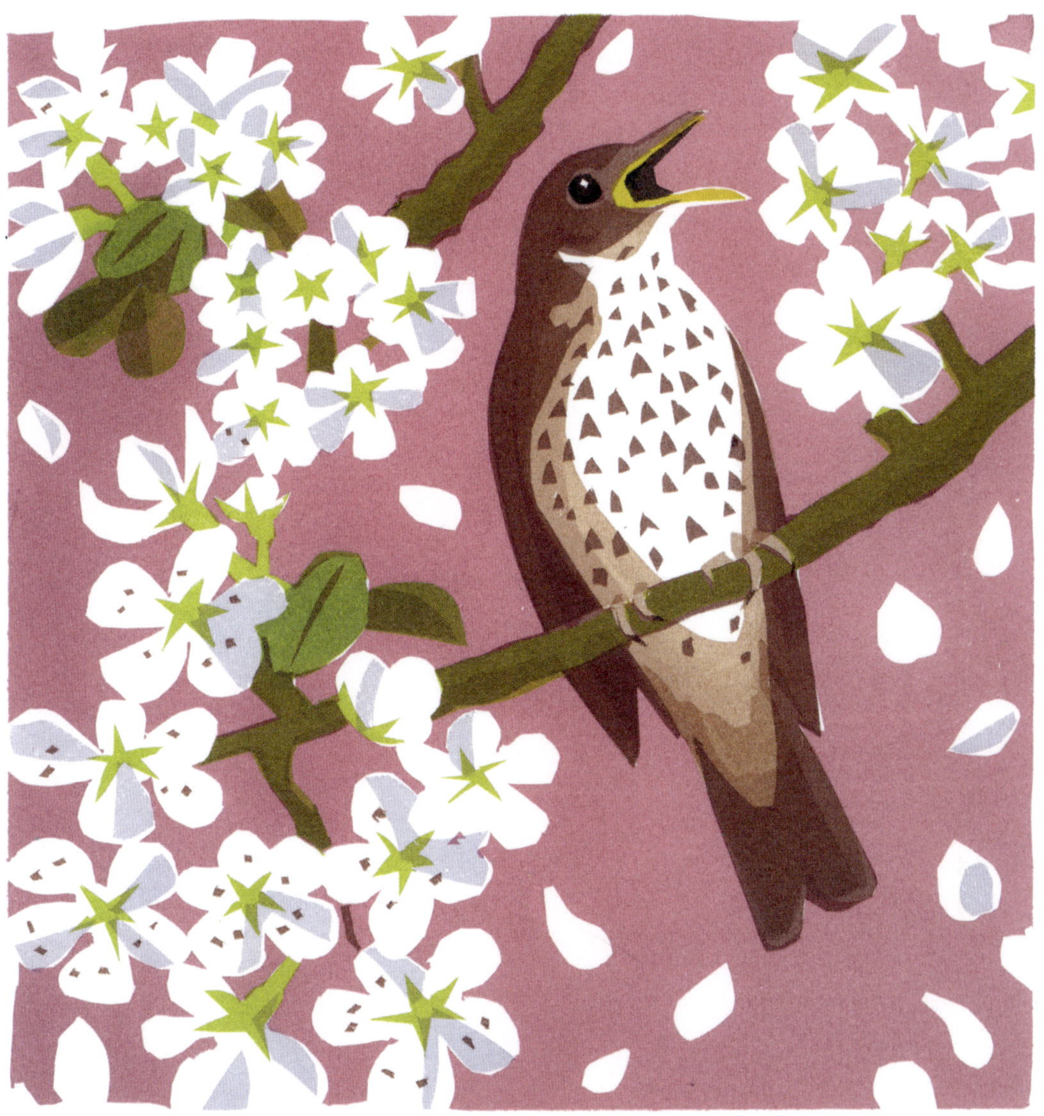

The **song thrush** (*Turdus philomelos*):

That's the wise thrush; he sings each song twice over,
Lest you think he never could recapture
The first fine careless rapture!

Robert Browning (1812–1889), from 'Home-thoughts, from Abroad'

A song thrush's repertoire can increase with age to 100 different phrases, meaning it can sing a million phrases in a year. It is also distinctive in choosing snails as its favourite food, cracking their shells against stones.

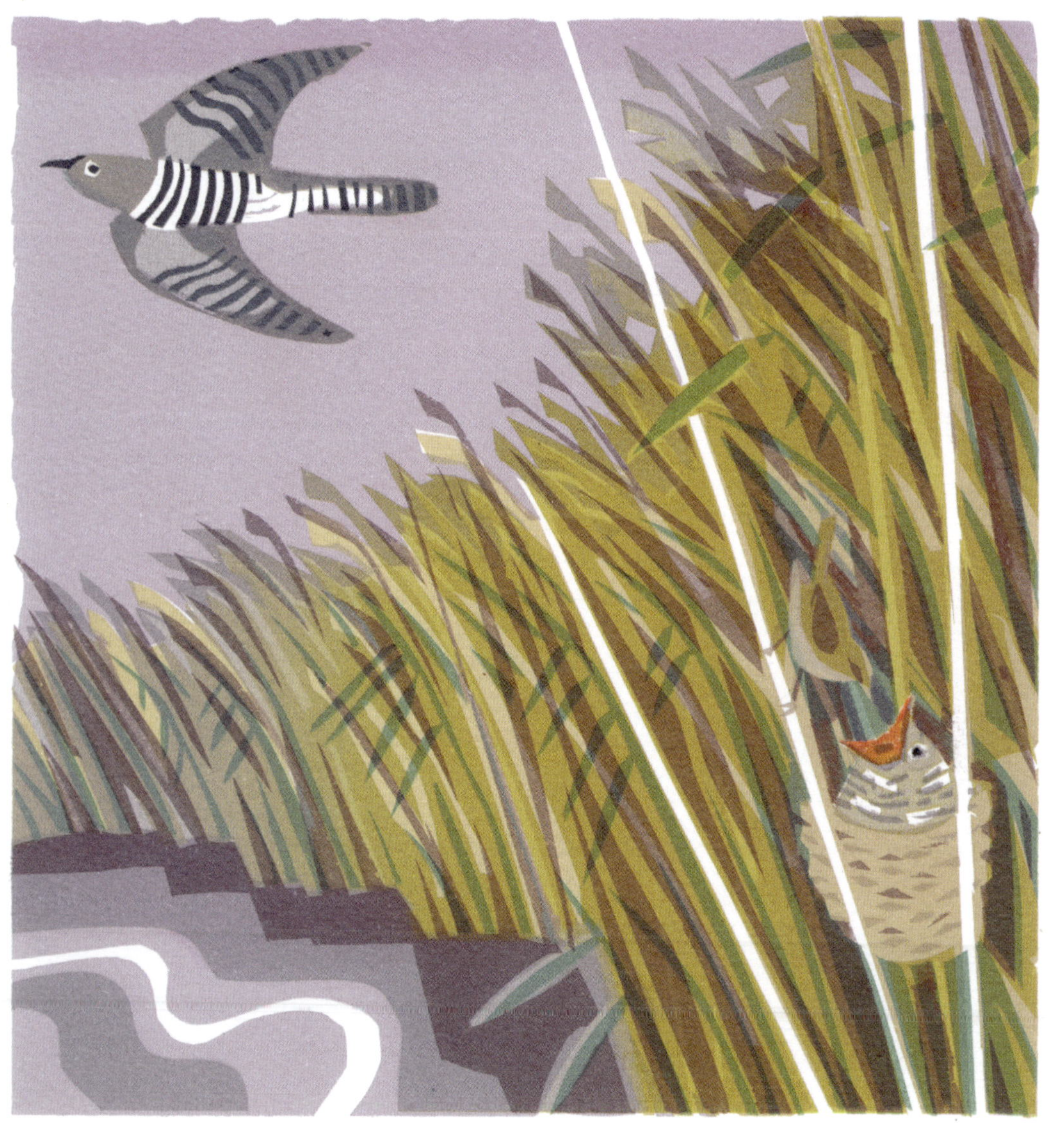

The anonymous thirteenth-century 'Cuckoo Song'

> *Sumer is icumen in*
> *Lhude sing cuccu*

is the oldest six-part polyphony in English music. The **cuckoo** (*Cuculus canorus*) is infamous for laying its egg in another bird's nest. In 1988, Mike Bayliss at the Trap Grounds nature reserve recorded a female planting 25 eggs, a world record, and 113 over eight breeding seasons. Three days after hatching, trebled in size, a cuckoo chick expels the foster bird's eggs or chicks. Fledged, parents unknown, it migrates to tropical Africa. A decline of 65 per cent since 1980 in the British Isles seems partly due to earlier springs in England.

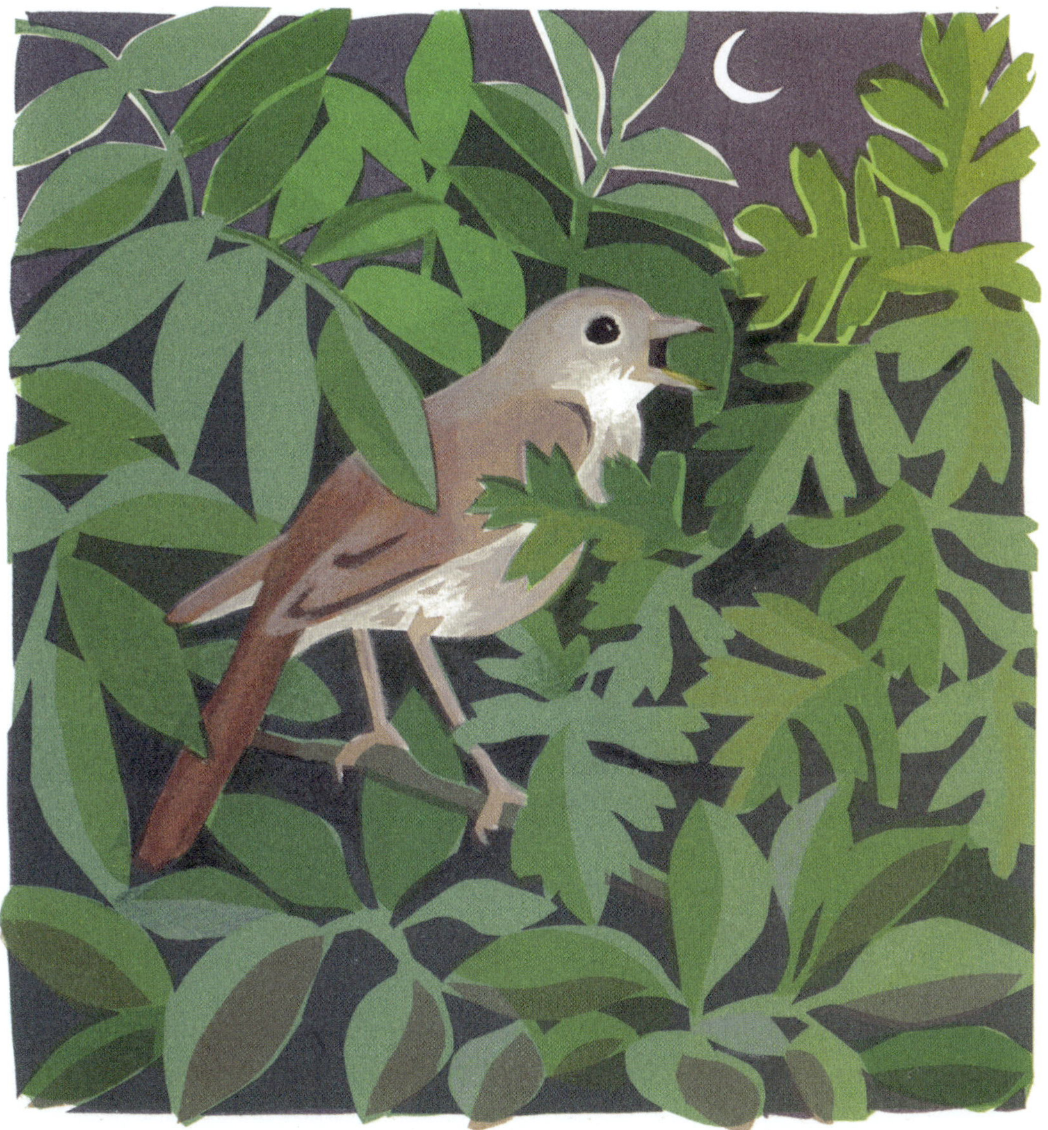

Because of its nocturnal singing, the **nightingale** (*Luscinia megarhynchos*) has inspired poets like no other bird. Yet, as John Clare wrote:

I could not think so plain a bird
Could sing so fine a song

From 'To the Nightingale'

Loss of habitat has seen it decline 90 per cent in 50 years. In 2003, farmers first received EU subsidies to create wilderness to benefit wildlife such as the nightingale. Isabella Tree and Charlie Burrell have achieved the best-known example of rewilding at Knepp Estate, Sussex, popularised by Tree's book, *Wilding*. The result – 2001: no nightingales; 2023: 50 singing males.

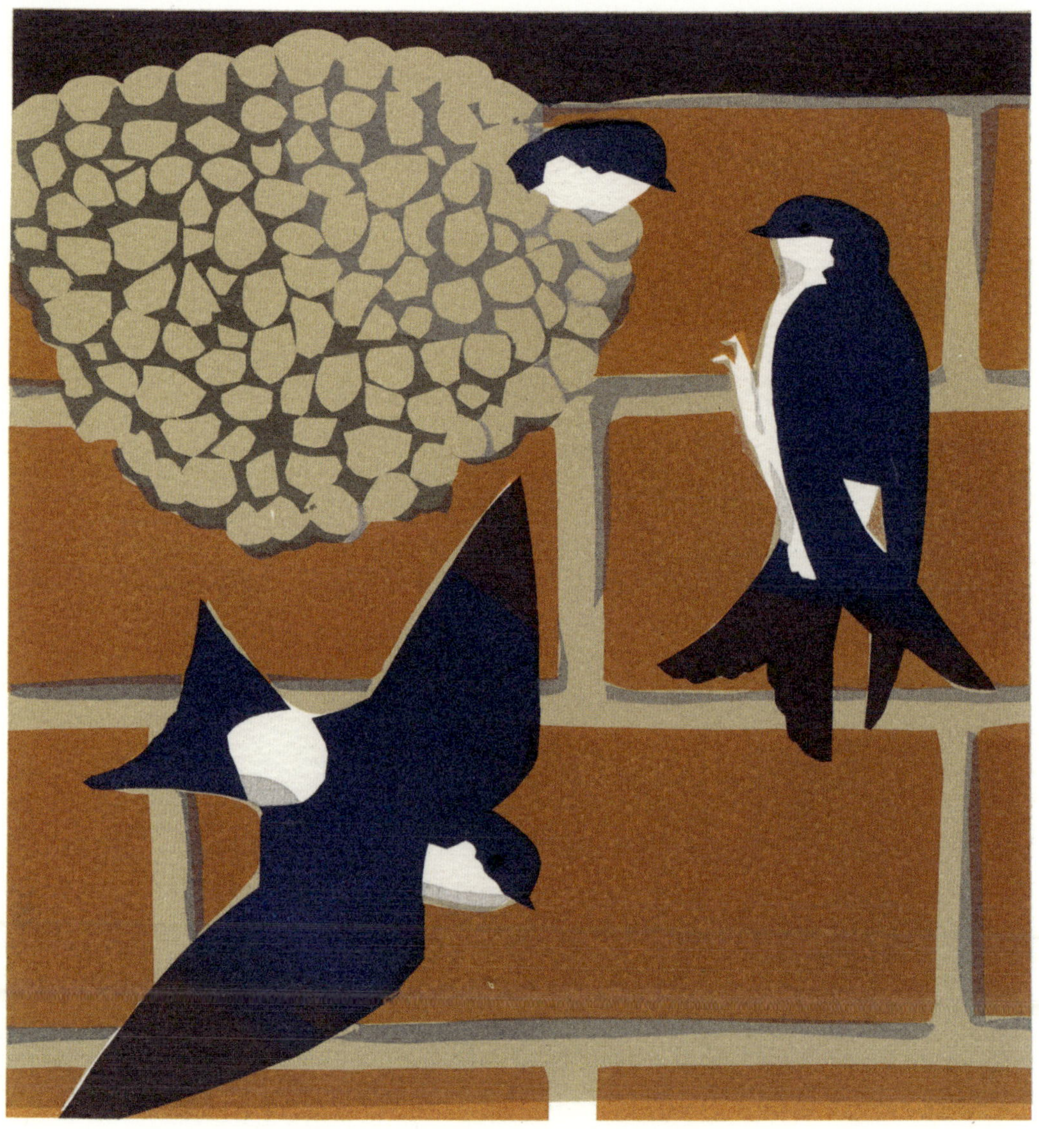

The **house martin** (*Delichon urbica*), the first bird to bear a Christian name (after St Martin), arrives in March and leaves by Martinmas, St Martin's feast day, on 11 November. Gilbert White liked the way house martins 'sport about', noting this left their morning-built mud nests to set in window corners through the day. In living memory a few nested on Lundy island's cliffs. They tend to have two or even three broods, which explains late departures. A colony nests on the facing embassies of France and Kuwait at Albert Gate, Knightsbridge, not far from Harrods and Hyde Park Corner.

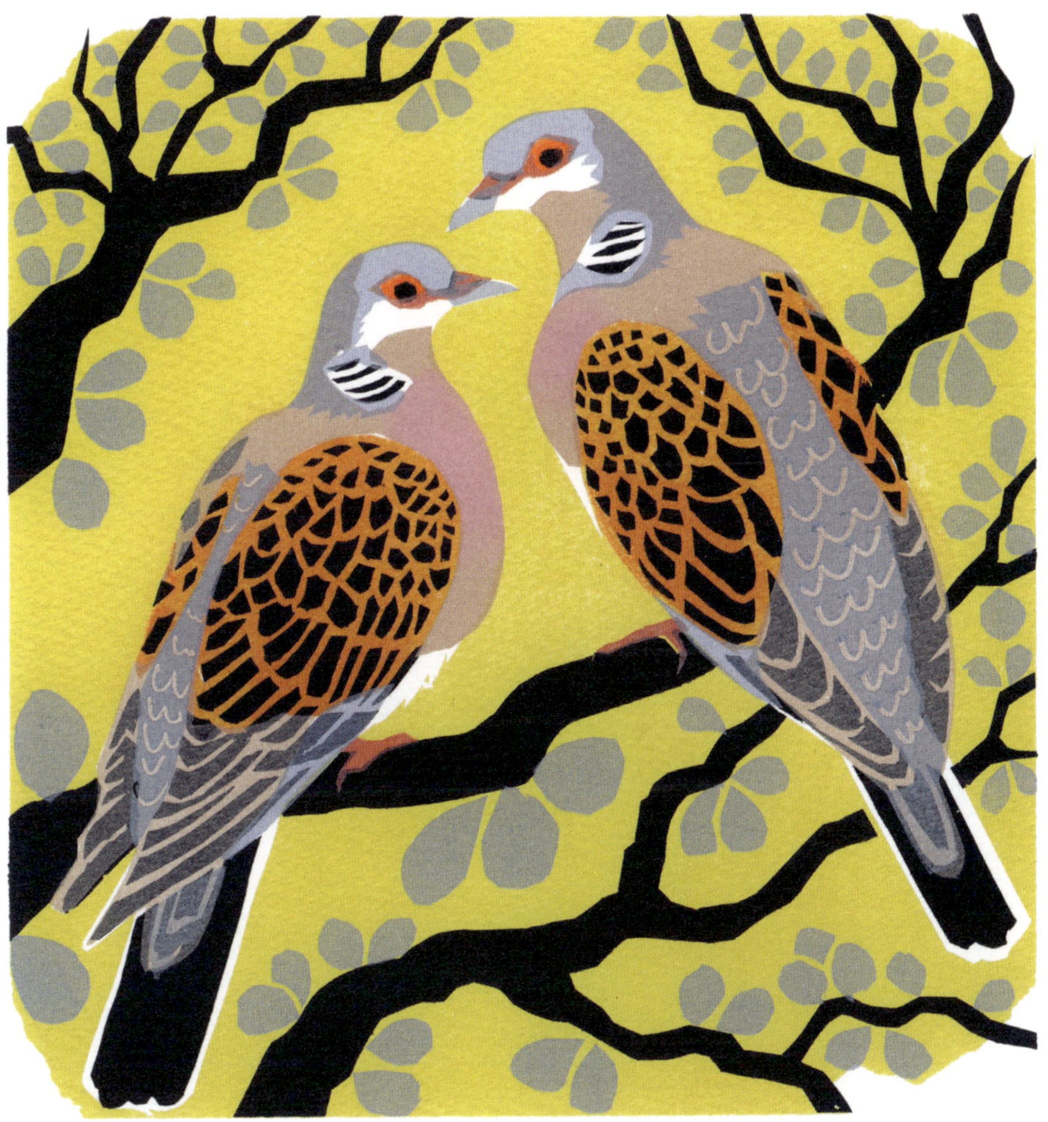

The mating fidelity of the **turtle dove** (*Streptopelia turtur*, from its *tur-turing* call) lends weight to 'lovey-dovey'. John Gay (1685–1732), from *The Beggar's Opera*, Air xiii:

The Turtle thus with plaintive Crying,
Laments her Dove.
Down she drops quite spent with Sighing,
Pair'd in Death, as pair'd in Love.

From the 1960s to the 1990s, annual clutches per pair of this beautiful migrant, with its soothing summer call, fell from 2.9 to 1.6; principal cause a lack of wild seed. UK pairs have declined from 125,000 (1970) to 2,200 (2022). Following Knepp's rewilding example, landowners are notable for attempting to join forces to stem the decline.

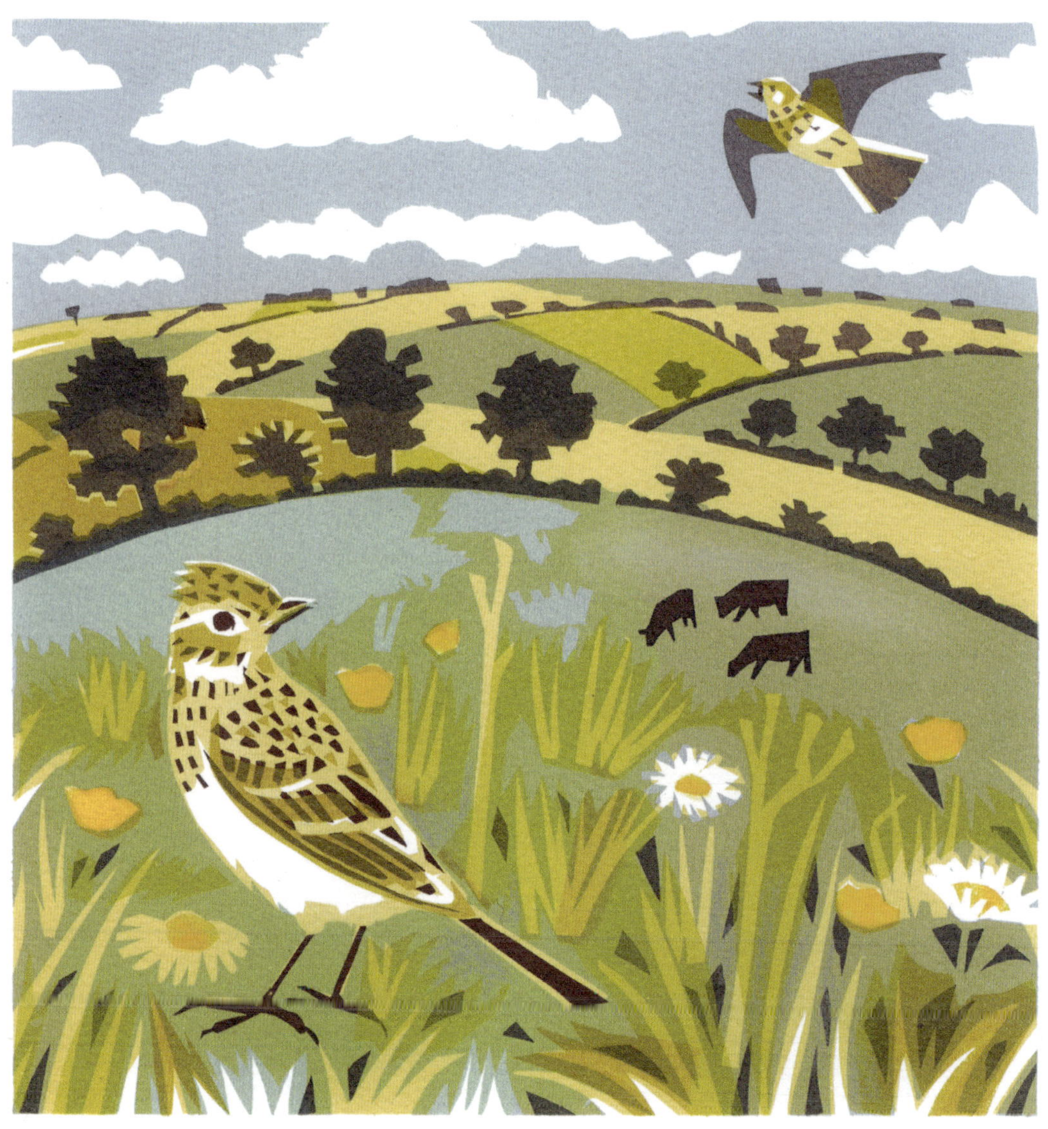

Higher still and higher
From the earth thou springest
Like a cloud of fire;
The deepest blue thou wingest,
And singing still dost soar and soaring ever singest.

Percy Bysshe Shelley (1792–1822), from 'To a Skylark'

The 460-syllable song of the **skylark** (***Alauda arvensis***) will transport it to 600m (2,000ft) (although it can deliver it stationary). In 2022 *The Lark Ascending*, the orchestral work by Vaughan Williams (1872–1958), topped the Classic FM 'Hall of Fame' poll for the 12th year running.

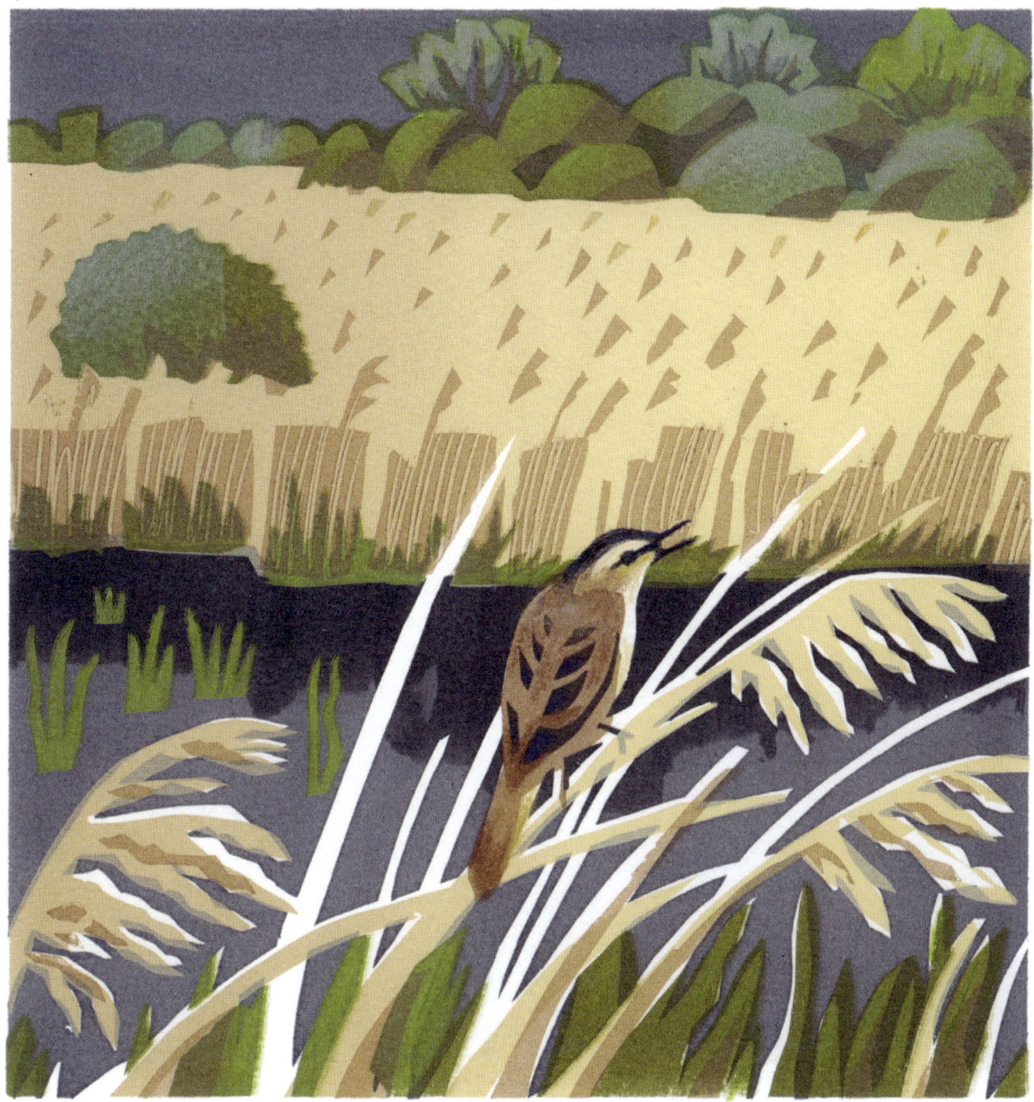

Gilbert White was first to distinguish the **sedge warbler** (*Acrocephalus schoenobaenus*) from the reed warbler 'by its creamy white eyebrow'. On 23 May 1915, poet Edward Thomas (1878–1917) strolled beside the River Meon, Hampshire. He wrote 'Sedge Warblers' the same day:

...And sedge warblers, clinging so light
To willow-twigs, sang longer than the lark
Quick, shrill or grating, a song to match the heat
Of the strong sun, nor less the water's cool...

Soon after, despite age and marriage exemption, he enlisted and was killed in action at Arras. Author and conservationist W. H. Hudson (1841–1922) mourned him as the son he never had.

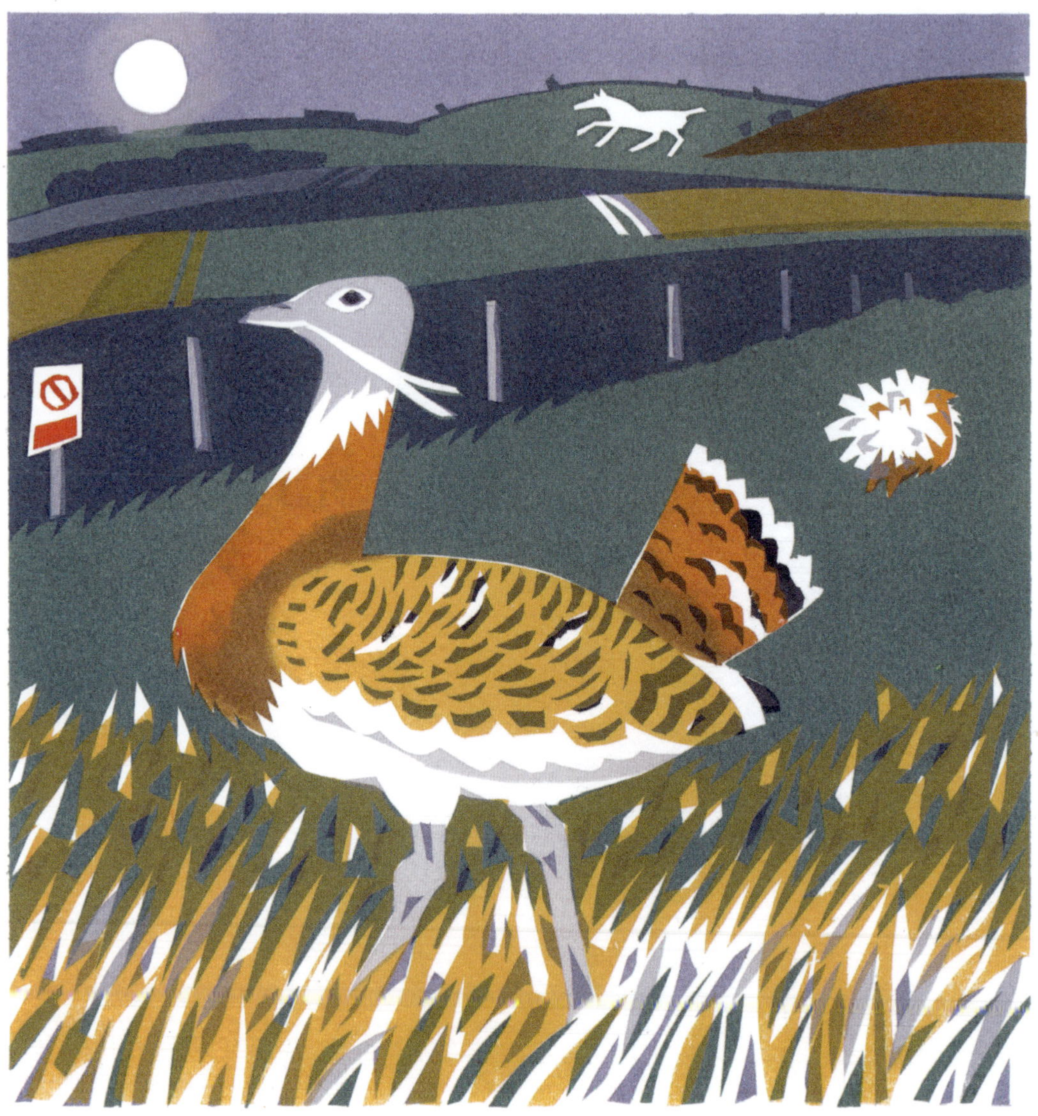

The **great bustard** (*Otis tarda*), along with the mute swan, is among the world's heaviest flying birds, with males of up to 18kg (39.5lb). It features in the coats of arms of Cambridgeshire and Wiltshire but has been extinct in the British Isles since 1832. Wildlife artists' gallerist Aylmer Tryon (1909–1996) set up the Great Bustard Trust in 1970, securing a 10-acre pen on Porton Down, Salisbury Plain. In 2004, the Trust's successor, the Great Bustard Group, released fledged birds parented in Russia onto Salisbury Plain. In 2023, there is a wild local population of 100 birds. Visitors to the site can see bustards flying and, if lucky, catch sight of a male's spectacular courting display.

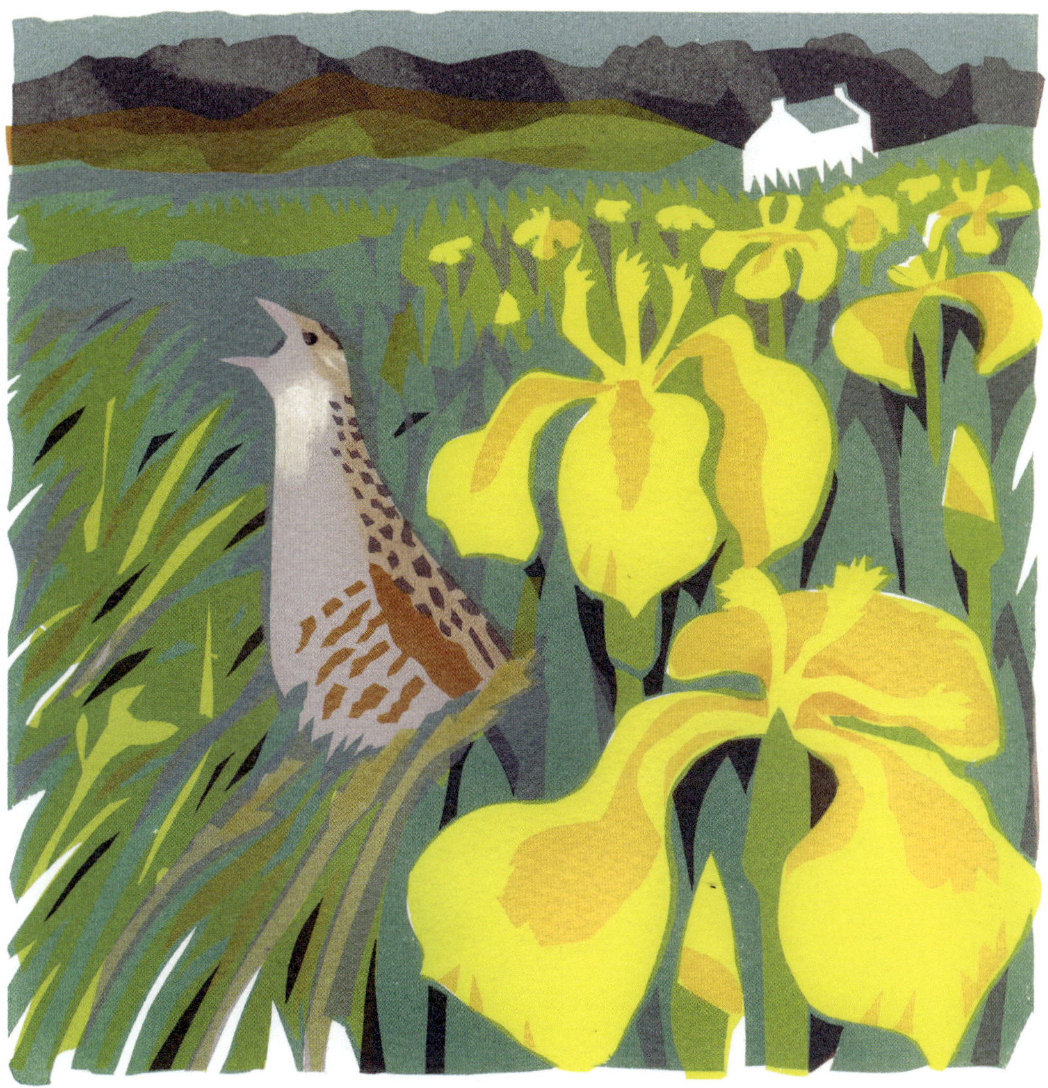

The scientific classification of the **corncrake** (*Crex crex*) echoes the male's rasping call, delivered day and night in May and June, yellow iris time. In 'The Landrail' (the species' old name), John Clare highlights how their call's reverberation disguises their location:

And still they hear the craiking sound
And still they wonder why
It surely cant be under ground
Nor is it in the sky

Earth-coloured, it prefers to hug the ground. Abundant 50 years ago, habitat loss has reduced UK numbers, but small RSPB reintroductions in England have been successful.

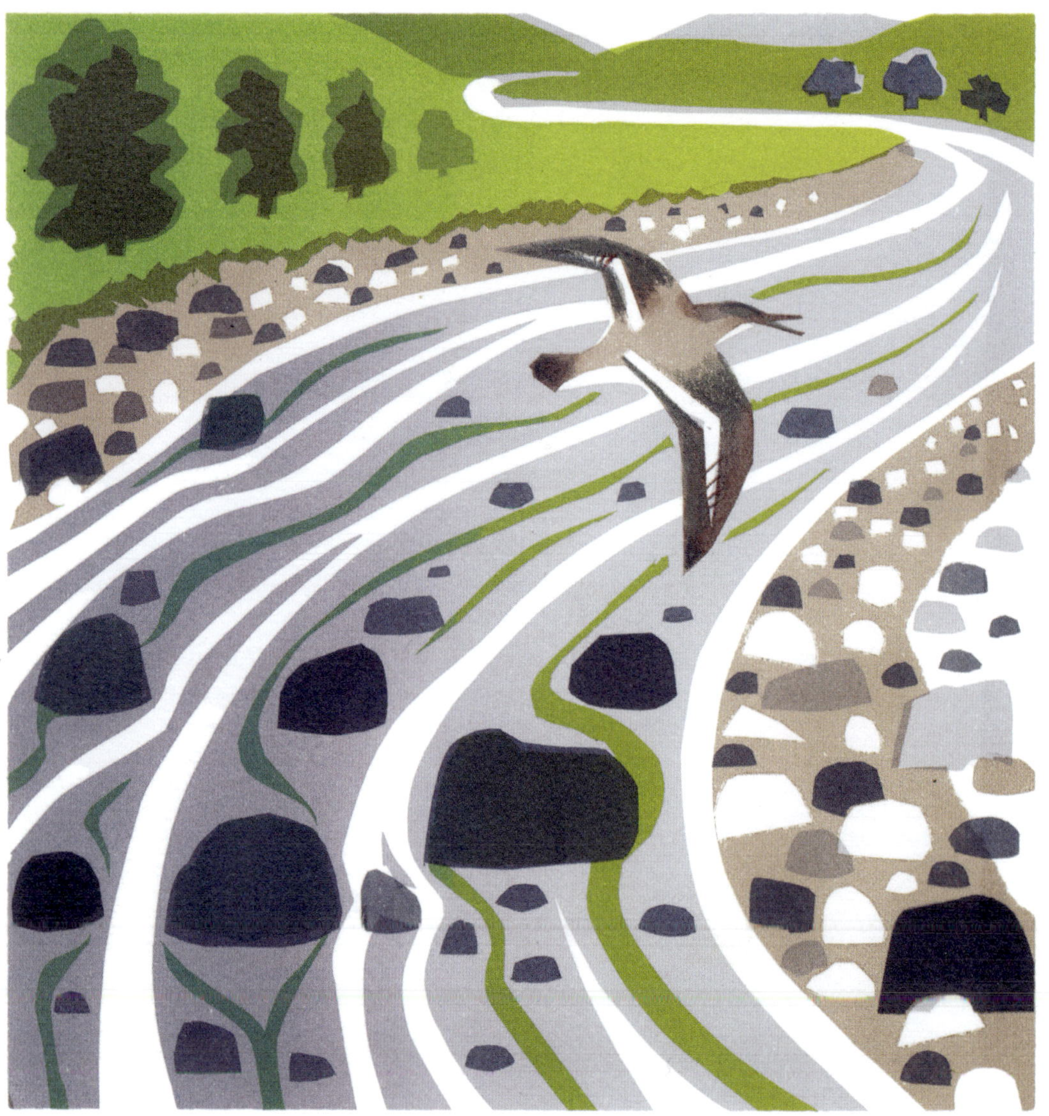

William Wordsworth (1770–1850), in his poem 'Lines left upon a Seat in a Yew-tree', refers to the **common sandpiper** (*Actitis hypoleucos*) as 'glancing'. And 'glancing' is the word, as it glimmers (*hypoleucos*, 'white beneath') across inland waters like a skimming stone in the pebble-throwing game of 'ducks and drakes'. A summer migrant from Africa, it is an electrifying addition to upland lakes and shingle-bordered rushing streams. Its pipe, high-pitched to be heard above the rapids, is haunting when a journeying pack passes in the night.

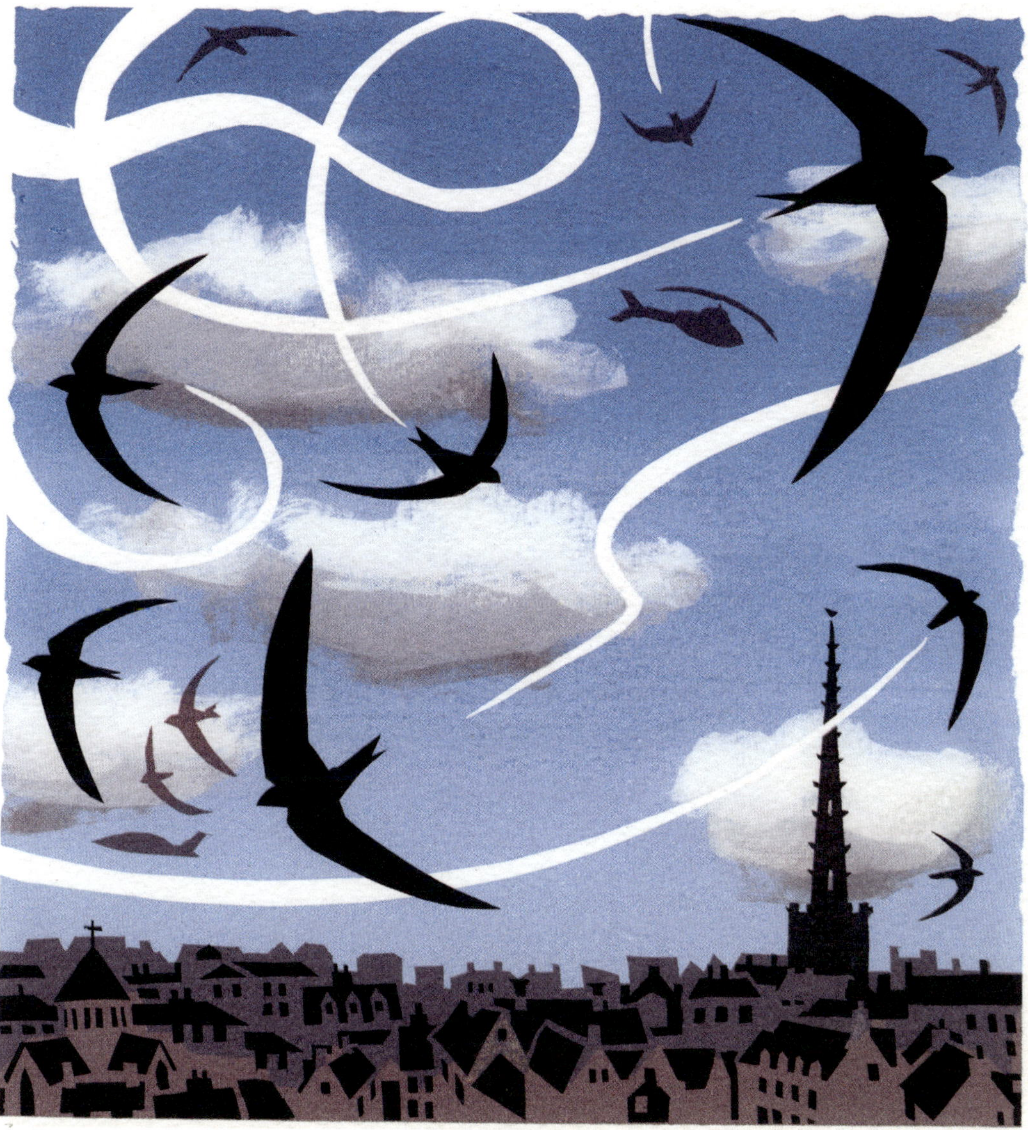

The solely insect-dependent **swift** (*Apus apus*) is the most aerial bird; with the longest wingspan at 76cm (30in) proportionate to its weight of 56g (2oz). It averages 805km (500 miles) a day, more than 3 million km in a lifetime. It eats, drinks and sleeps on the wing, reaching nocturnal heights of 3.2km (2 miles). In sleep, its unihemispheric state means one-half of its brain is alternately asleep or awake, allowing it to hold its course even on migration. Nestlings survive through dormancy if the weather deprives them of food. The modernisation of old buildings has reduced nesting options, one cause of a 60 per cent UK decline since 1995. It is hoped a swift brick for nesting will soon be mandatory for new UK houses.

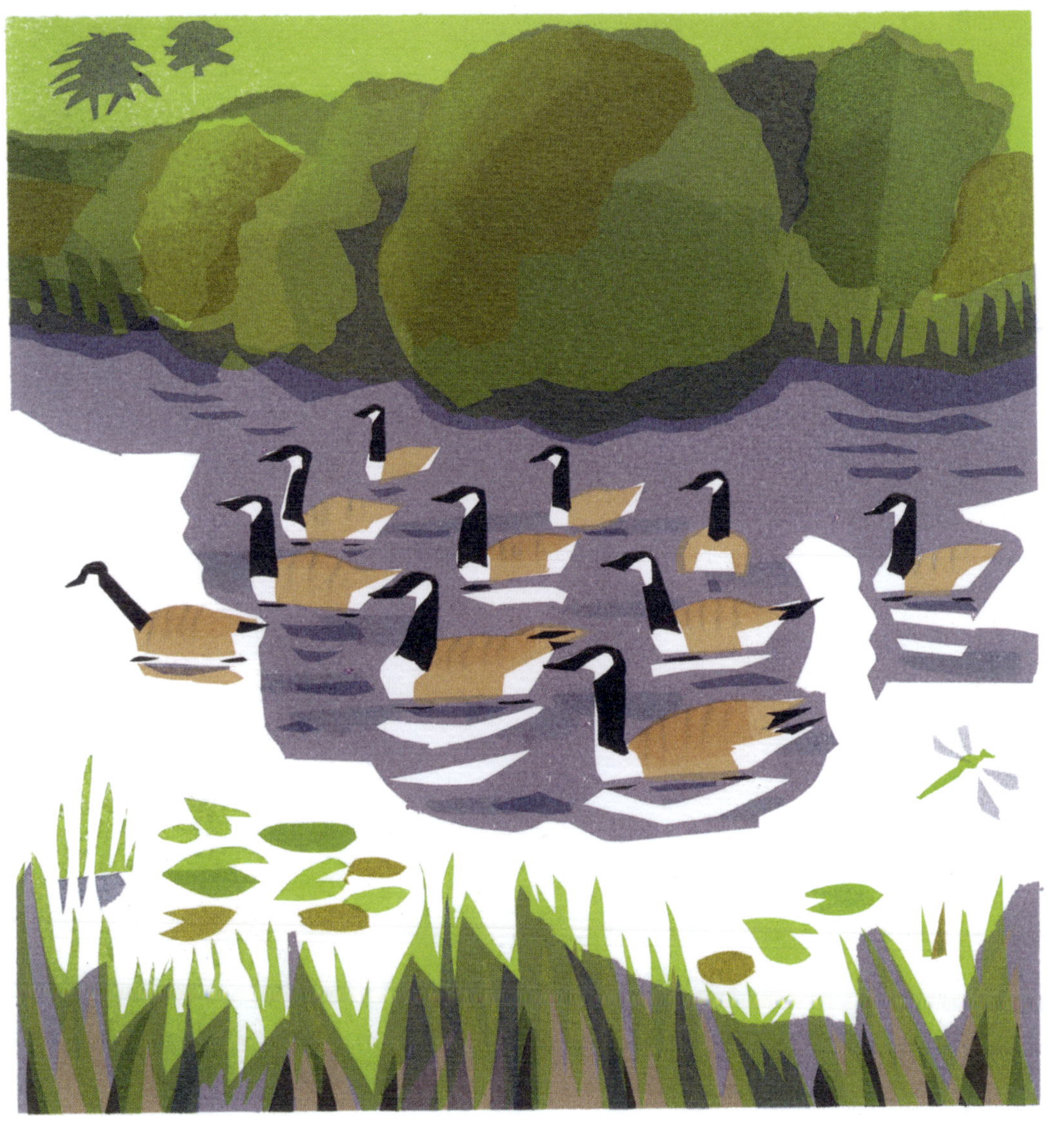

The **Canada goose** (*Branta canadensis*) was introduced from its native North America in the reign of Charles II for display in St James's Park. Today, at 54,000 resident UK pairs, it frequents municipal parks and is a public nuisance; it is no longer protected on private land. Its grazing consumption equals that of a sheep, and every 40 seconds it defecates, droppings amounting to 1kg per day and carrying superbugs resistant to antibiotics. Airborne, its skeins have caused one fatal airline crash in Canada and forced a US Airways plane to land on the Hudson River off Manhattan in 2009.

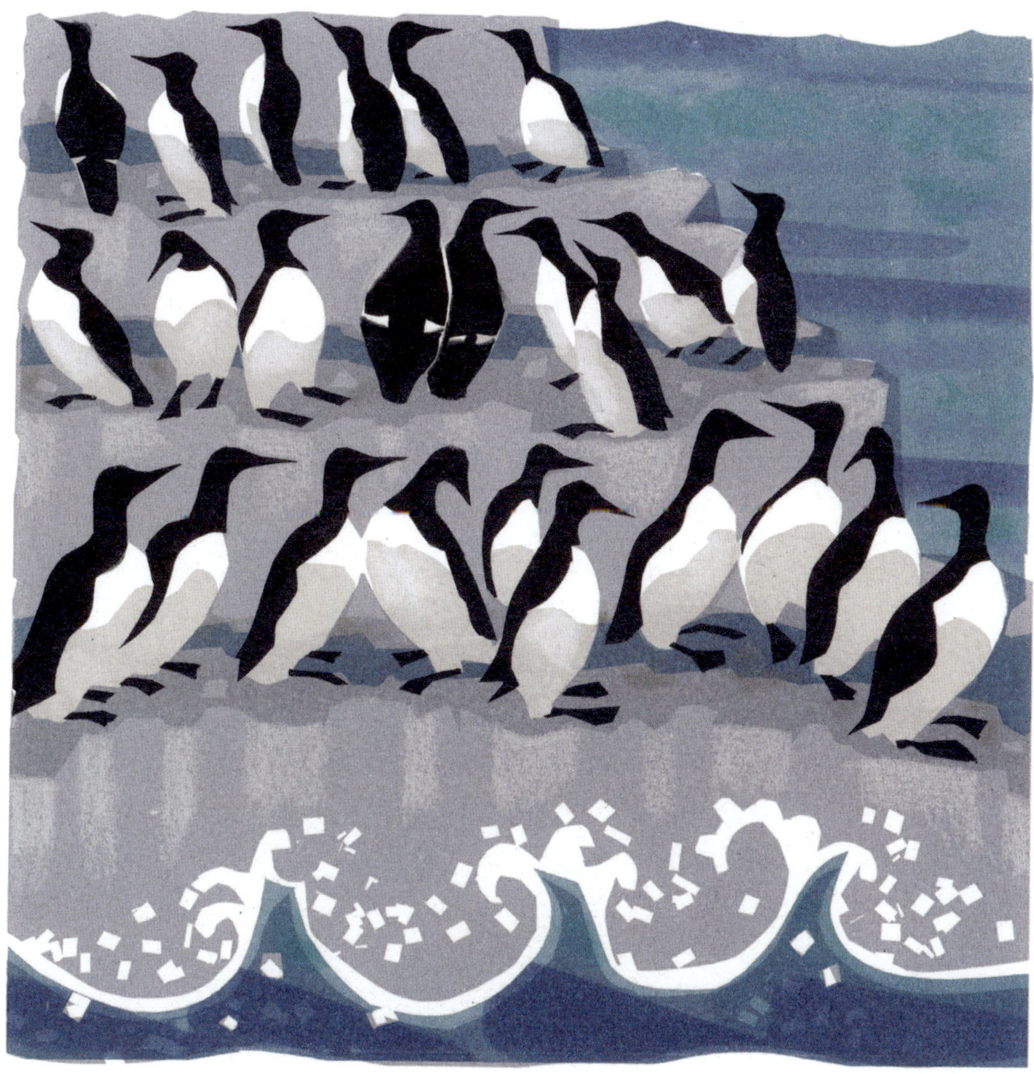

Collecting wild bird eggs has been illegal since 1954. As a food source and aesthetic pursuit it had carried on for centuries, but its aesthetic zenith was the 1930s at Bempton Cliffs on the Flamborough headland, East Yorkshire. Collectors and local 'climmers' (climbers) bartered at the cliff edge for the eggs of the **guillemot** (*Uria aalge*), the world's ultimate bird egg, each unique in colour and pattern. The tapered end of the egg becomes mucky in the wild, the chick emerging from the cleaner oval end. The Koh–i-Noor diamond of bird eggs is the Bempton Belle, a red guillemot egg found by J. W. Goodall in Yorkshire. The 1,000-strong guillemot egg collection of oologist George Lupton (1881–1970) is in the Natural History Museum, Tring.

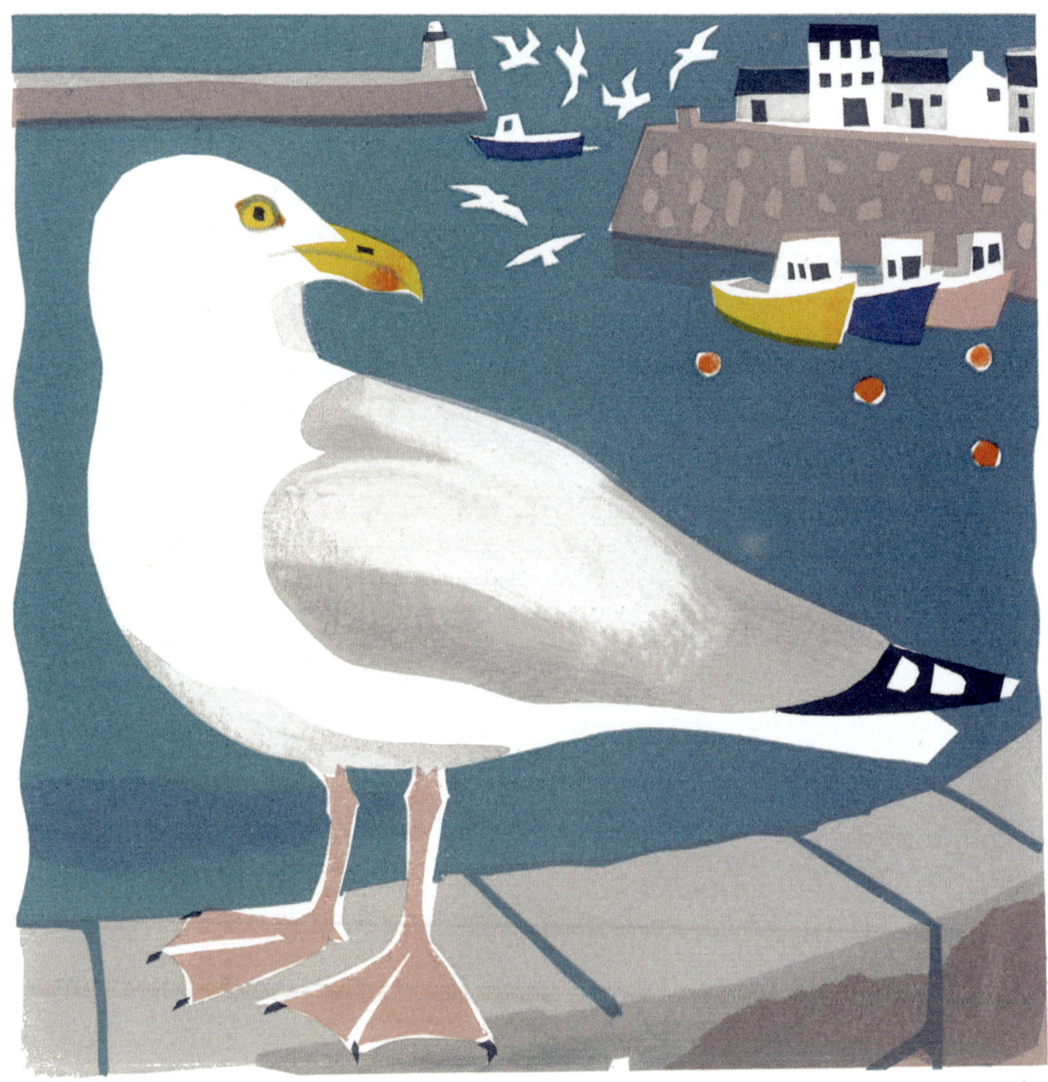

When fishing ports thrived, the **herring gull** (*Larus argentatus*) fed on fish and nested on cliffs. In the 1960s, with fish stocks declining and fast food proliferating, many became urbanised, scavenging garbage and nesting on roofs. In 2023, Blackpool Zoo employed volunteers to don inflatable eagle costumes to scare the gulls away. Concern the UK's resident pairs are falling below six figures has placed it on the red 'species of conservation concern' list, but half a million or so individual migrants boost that figure in winter. The British media continues to sensationalise its misdeeds at seaside holiday time. Nonetheless, herring gulls still romantically symbolise the sea. By popular demand their oceanic cries introduce BBC Radio 4's *Desert Island Discs*, first recorded in 1942 and aired most weeks since.

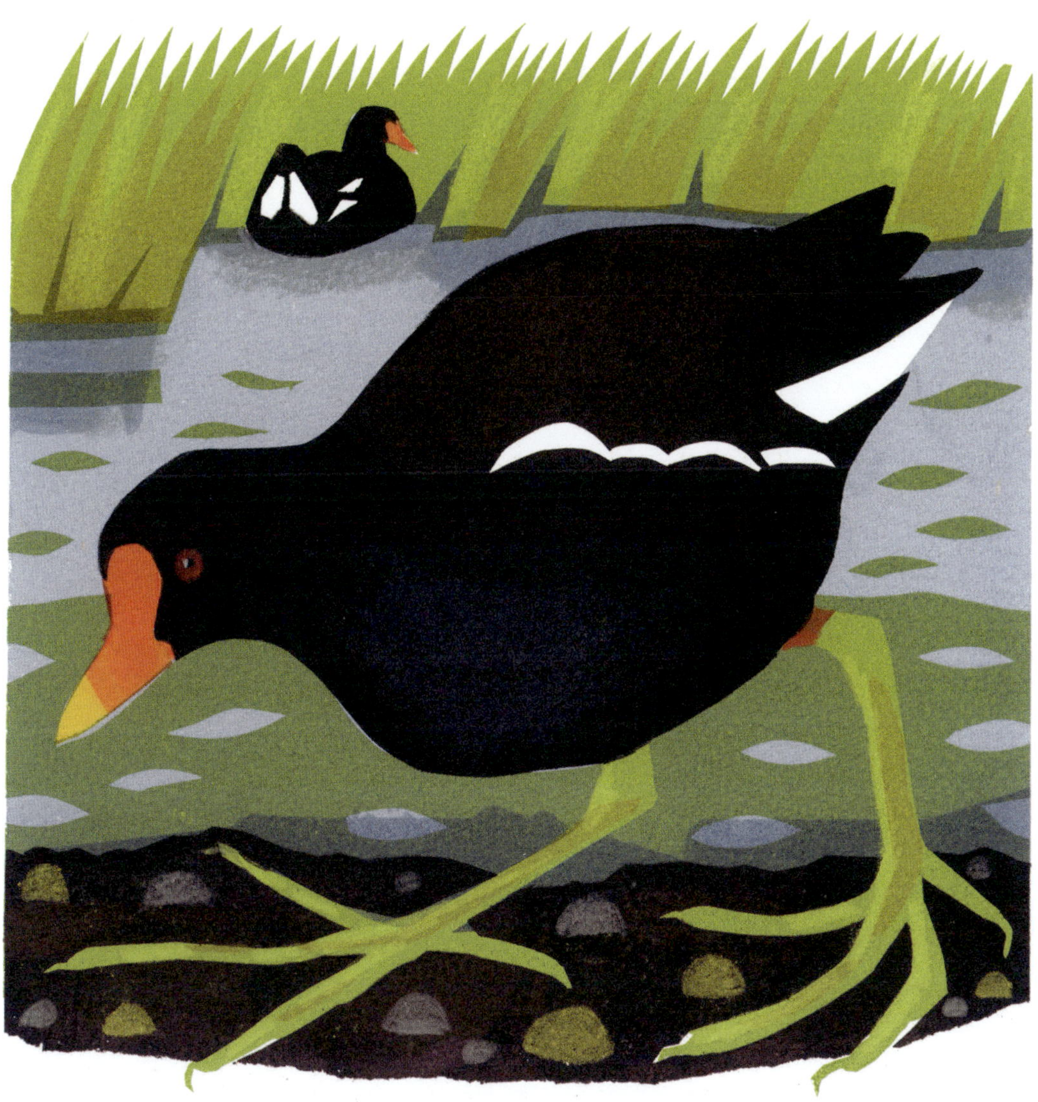

'Mere-hen', morphed into **moorhen** (*Gallinula chloropus*, also 'water hen'), is 'the best-behaved bird on the river', according to Adrian Richmond (see feral pigeon), also an experienced angler. The adult's white tail under-feathers act as a flag to the chicks. The moorhen's feet could fit a much larger bird, but are ideal for gripping water-hugging branches on which, for security, it often builds its nest:

...though danger comes
It dares and tries and cannot reach their homes
And so they hatch their eggs and sweetly dream
On their shelfed nests that bridge the gulphy stream

John Clare, 'The Moorhens Nest'

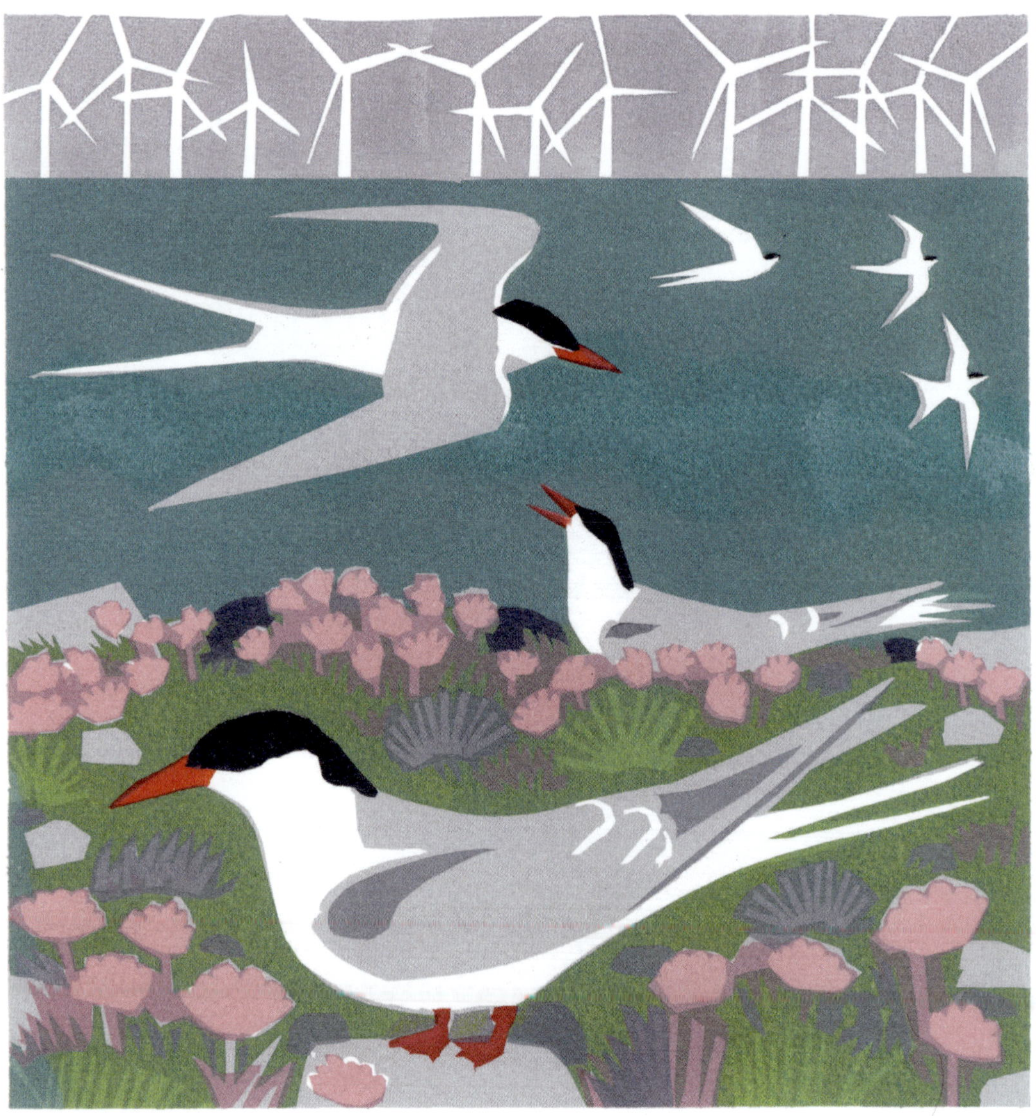

The **Arctic tern** (*Sterna paradisaea*) migrates annually from Pole to Pole, the longest journey of any creature at 90,123km (56,000 miles), a quarter of the way to the moon. It weighs 113g (3.9oz), with a forked tail and a 76cm (2.5ft) wingspan, its body propelled by little more than a heart held by wing muscle. In spring and autumn passage it can be seen over town lakes, its graceful undulating flight as it circles and dives for fish making it immediately distinguishable from a gull. There is another fork-tailed summer migrant 'sea swallow', the common tern. It has a black tip to its similarly scarlet bill. Birders hedge their identification bets with 'comic tern'.

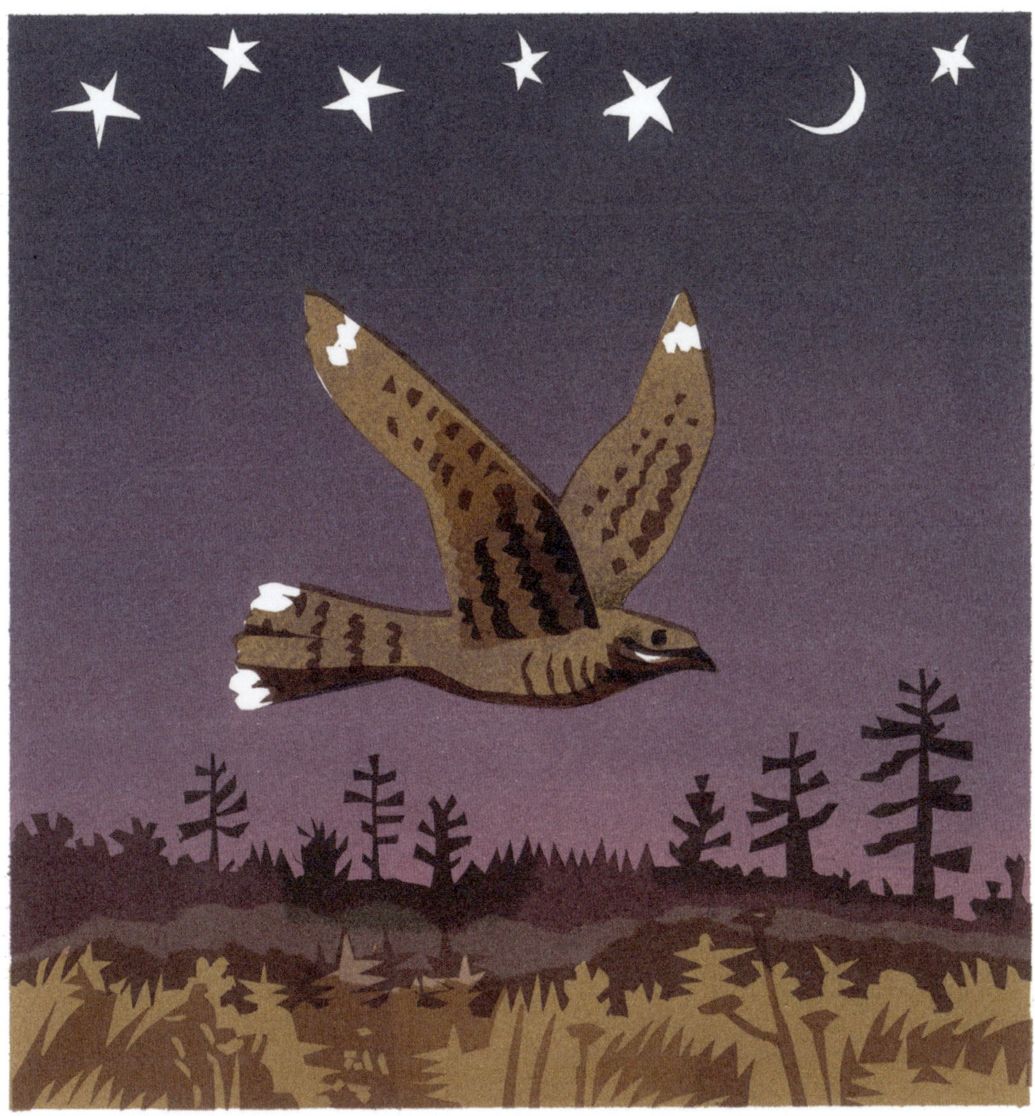

A favourite terrain of the summer migrant **nightjar** (*Caprimulgus europaeus*) is a heath, many of which have become airports (such as Heathrow). In the absence of a heath, a likely place to hear its twilit 'churring', as it hunts for moths and other insects in the falling dark, is the edge of a wood – best when undergrowth is rich after a recent felling. By day they are only conspicuous when sunbathing on a lonely road, hawklike as they fly off. Plumage offers such camouflage, their 'nest' is on bare ground.

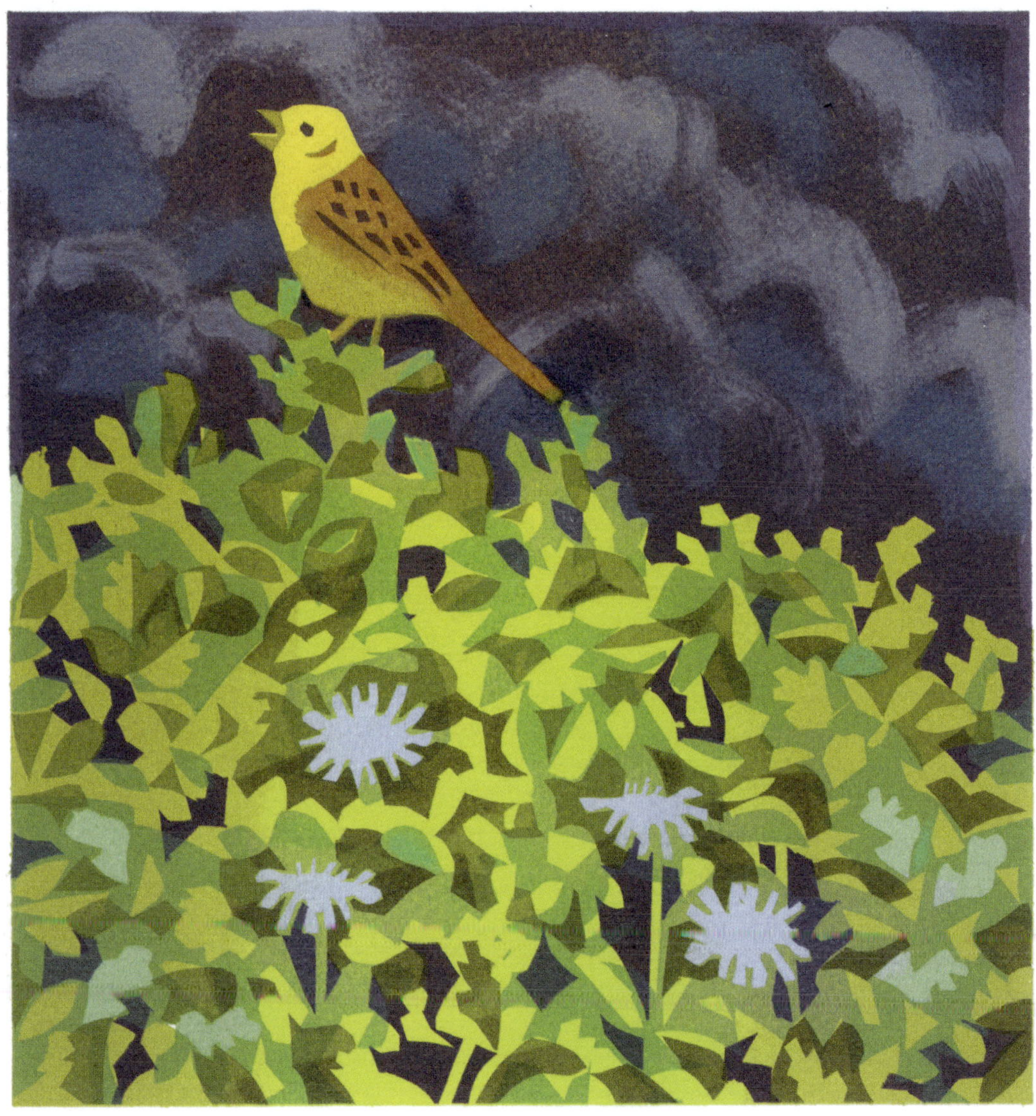

The male yellowhammer (*Emberiza citrinella*), of the bunting family, is one of the few birds to continue singing into autumn. Its song, transcribed as 'a-little-bit-of-bread-and -no-cheese', used to be the best-known mnemonic after 'cuckoo'. Both are rarer today, although less so in the case of the yellowhammer (700,000 UK territories, 2016). One of its names was 'scribble lark', which describes its egg pattern, the songbird equivalent of the guillemot's. John Clare's descripton could fit both, albeit the guillemot has only one egg.

Five eggs pen-scribbled over lilac shells
Resembling writing scrawls which fancy reads

From 'The Yellow Hammers Nest'

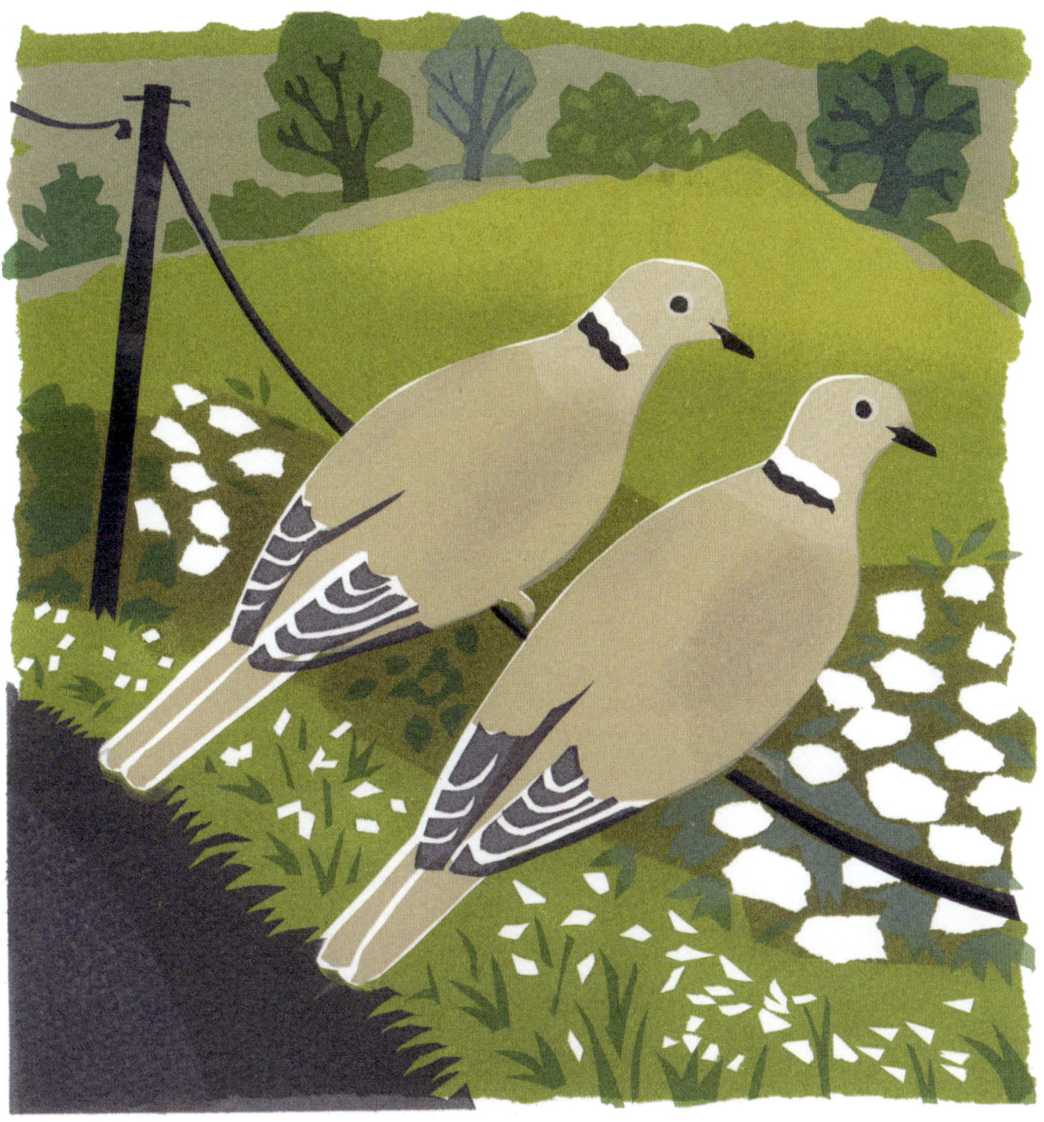

The **collared dove** (*Streptopelia decaocto*) first nested in Great Britain at Cromer, Norfolk, in 1955. Today it numbers 810,000 UK pairs. The expansion from Asia into Europe began in earnest in the 1930s. Its adoption of town and country and capacity to breed throughout the year explain its success. *Decaocto*, '18 coins', refers to a Greek myth of a servant girl who prayed to the gods for justice because of her pittance of a wage. Zeus answered her plea by creating the collared dove, so its ubiquity and mournful call would act as a perpetual reproach to stingy employers.

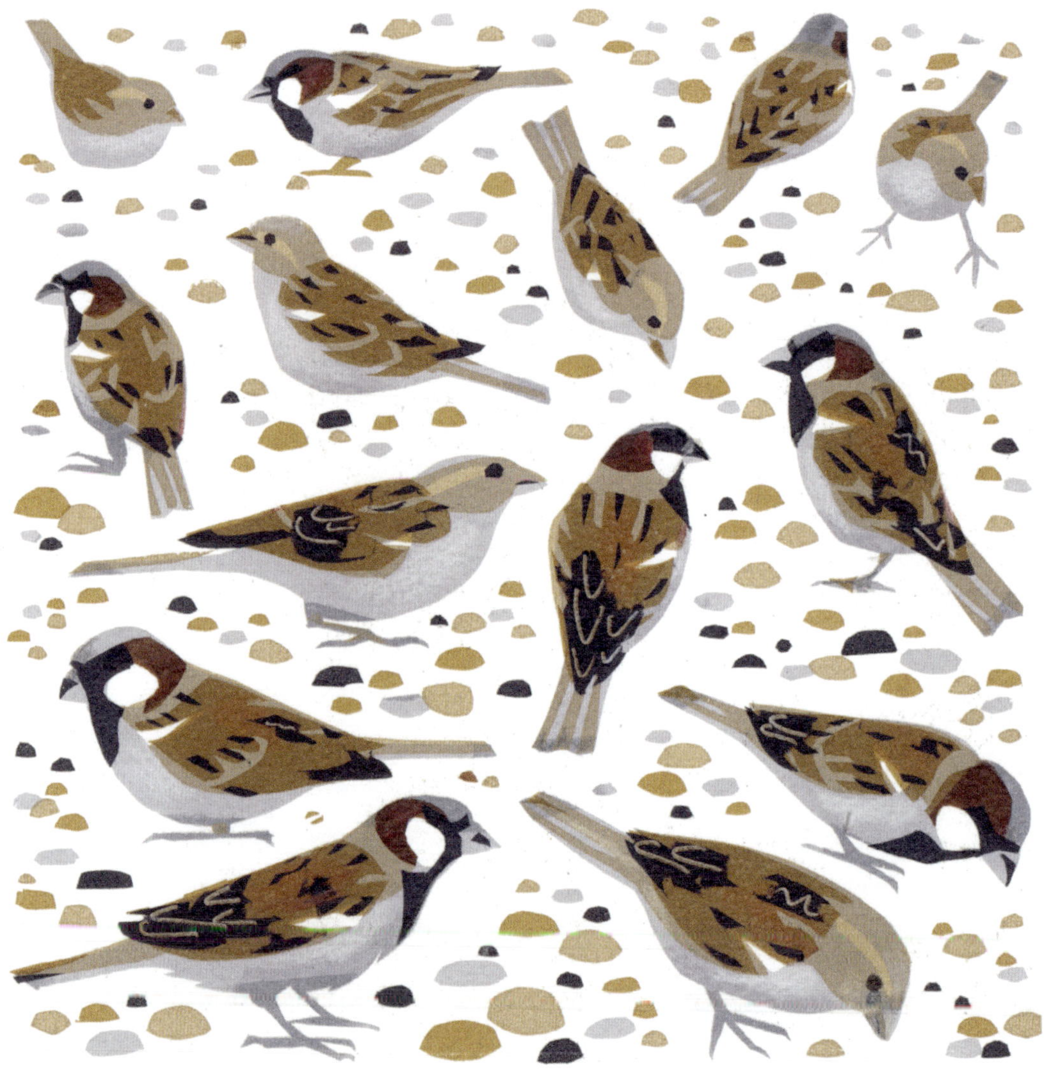

For Aristotle, the **house sparrow** (*Passer domesticus*) was 'of all birds most wanton'. The absence of its chirruping in urban areas makes its UK decline (from 30m in 1977 to 10m by 2016) particularly noticeable. Max Nicholson introduced the idea of a bird census to Britain in 1925 by counting the 2,603 sparrows in Kensington Gardens. He last counted in 2000, finding only eight. The Venerable Bede (690–735) described a sparrow's unseen flight over banqueting revellers, then out again into the night, as a metaphor for life's chilling brevity. Bede wrote the first history of England and popularised the term AD (Anno Domini: 'In the year of the Lord').

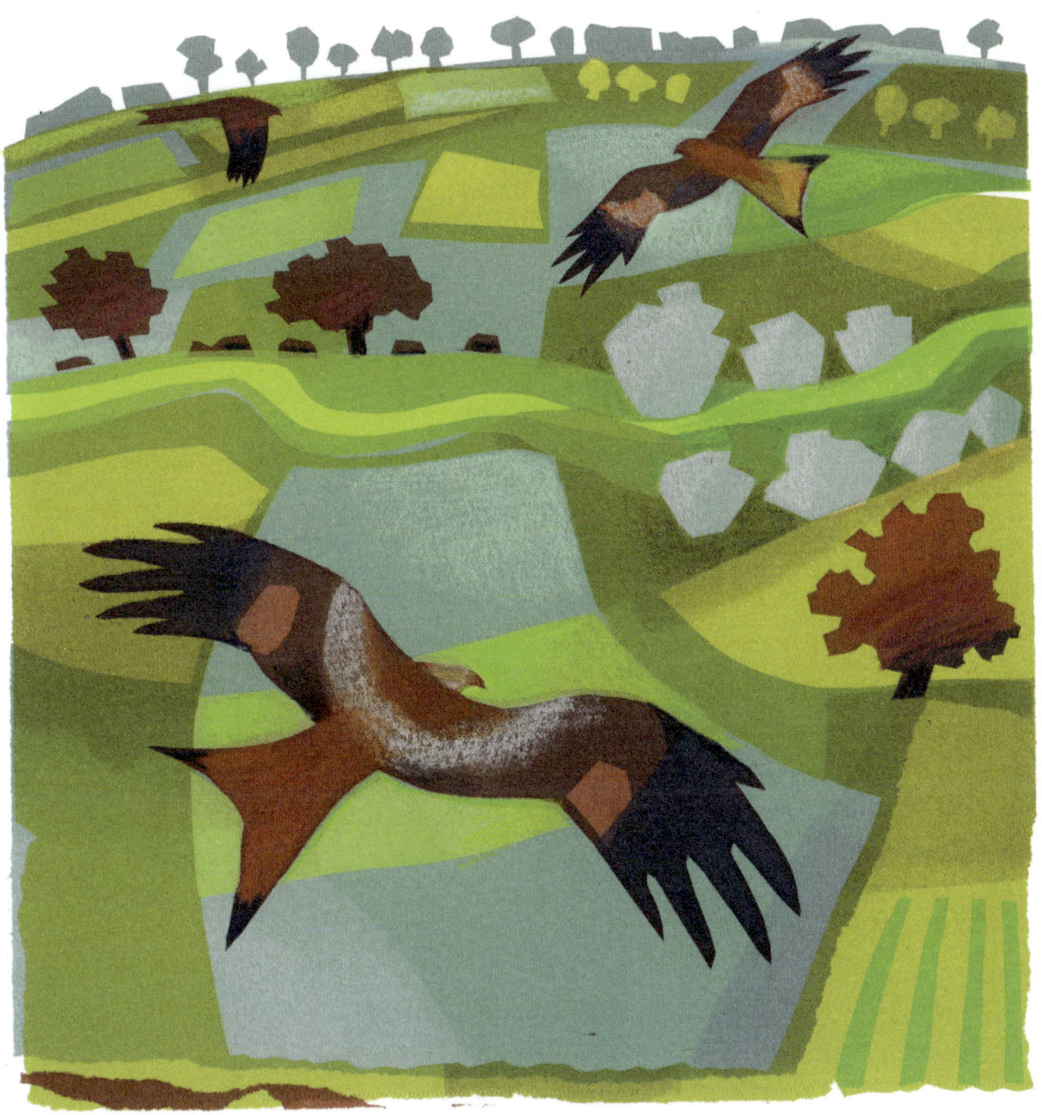

In 1951, the **red kite** (*Milvus milvus*) in the UK was reduced to six pairs. In 1989, Sir Paul Getty (1931–2003), a saviour of *The Oldie*, underwrote an RSPB introduction of red kites to England on his Buckinghamshire estate. A UK population of 50,000 birds (from 4,400 pairs in 2016) is predicted within years, to the despair of Richard Ingrams, founding *Oldie* editor, who commissioned and named the 'Bird of the Month' feature. 'They're too big', he wrote, 'especially when seen in numbers all the time.' Wales remains the heartland, with Rhayader proclaimed 'home of the red kite' and Tregaron's Kite Centre and Museum. Mentioned in the diary of Samuel Pepys (1633–1703), their taste for rubbish may herald a return to city centres.

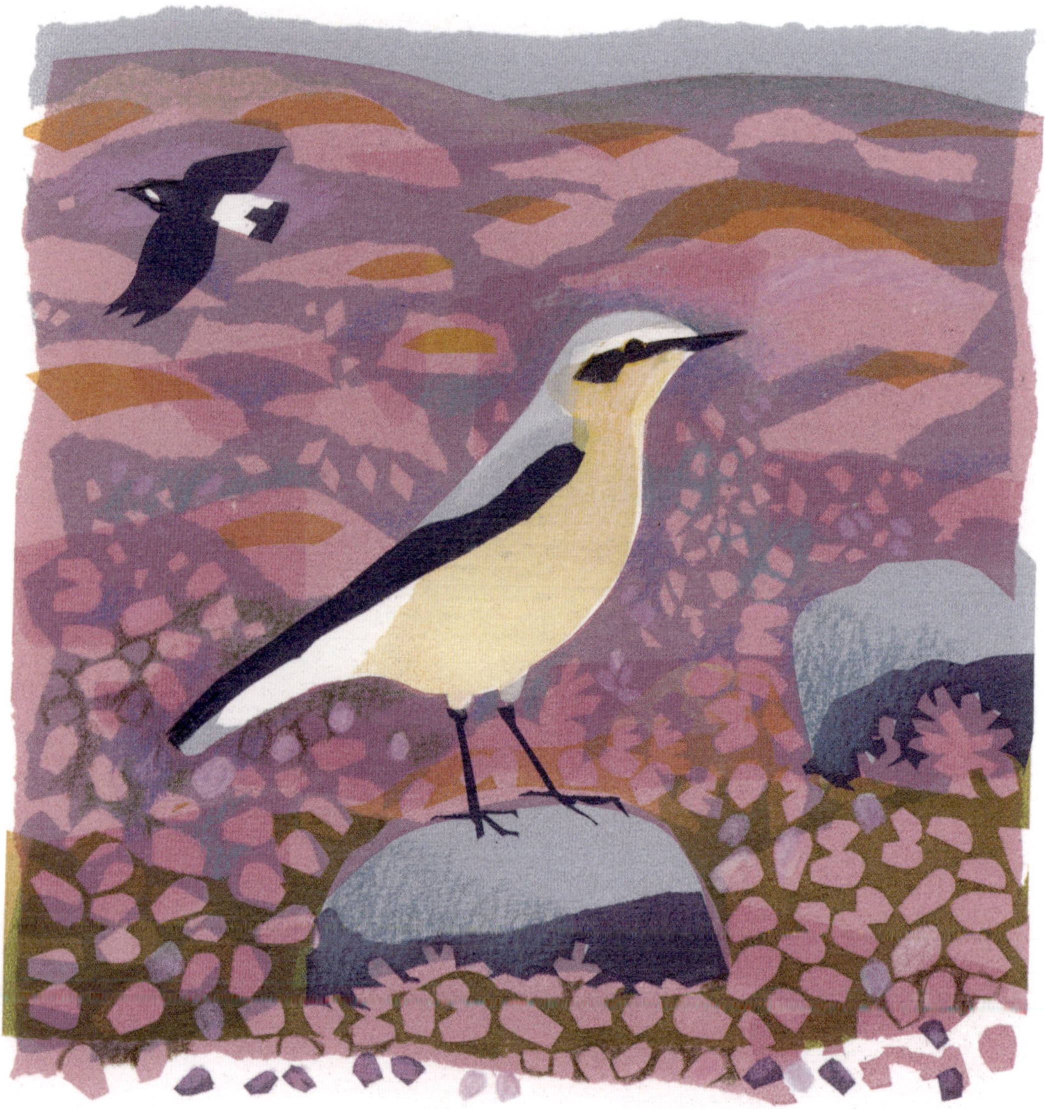

'To those who are attracted to solitary, desert, places, who find in wilderness a charm superior to all others, the **wheatear** [*Oenanthe oenanthe*] ... is a familiar figure – a pretty little wild friend; for he, too, prefers the uncultivated wastes, the vast downs, the mountain slopes and the stony barren uplands,' wrote W. H. Hudson (*British Birds*). *Oenos* is Greek for wine, *anthos* for flower; the wheatear's arrival in Greece is celebrated as the grapevine blooms. This gives scant indication of its migratory range from sub-Saharan Africa, the birds splitting east and west with some heading to North America and others to northern Europe.

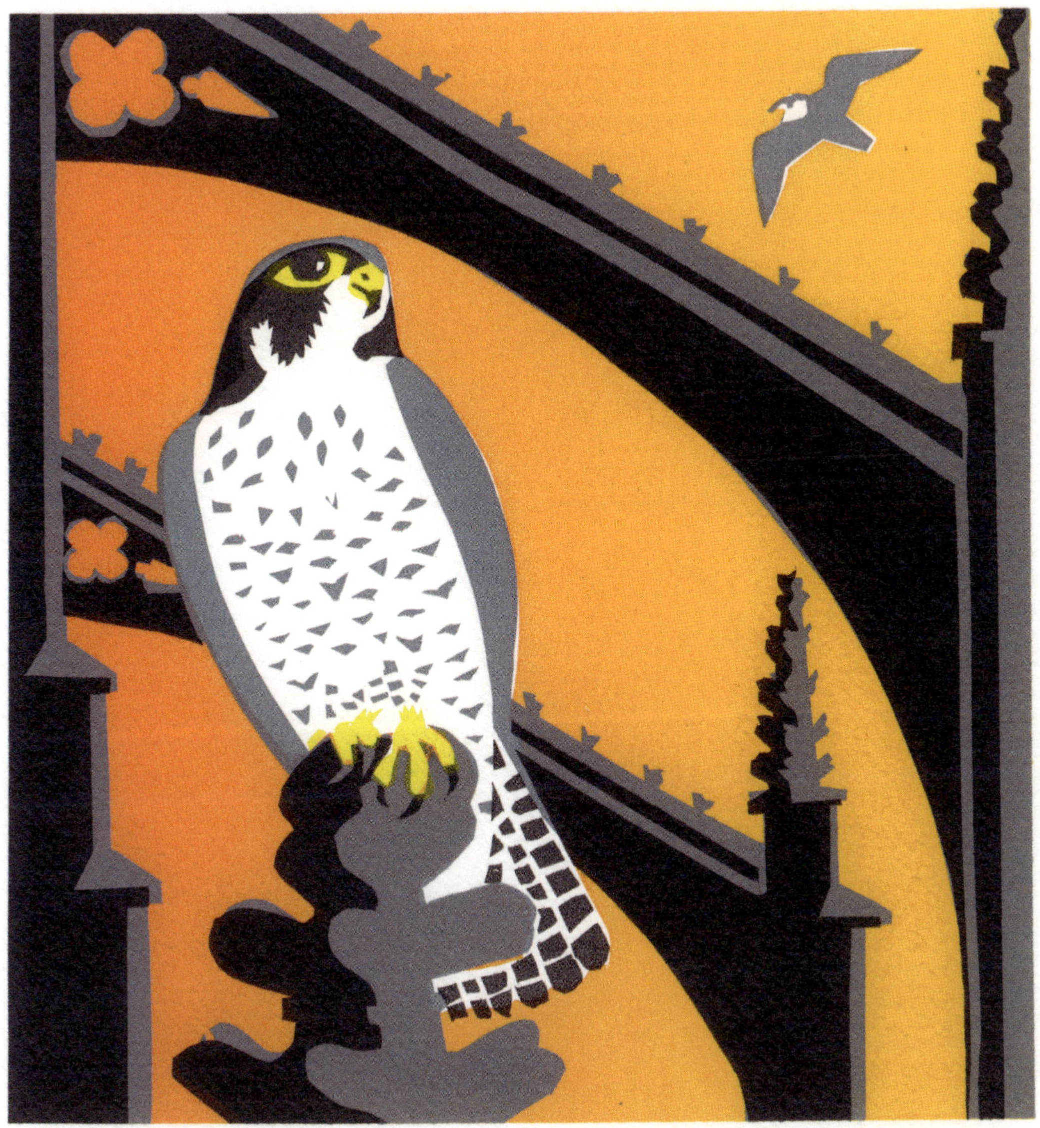

The **peregrine falcon** (*Falco peregrinus*), the world's fastest animal, can stoop at a record 389km/h (242mph). Thanks to legal protection and pesticide controls, especially of DDT which caused its eggs to break, the UK population has now reached 1,500 pairs. The peregrine's mewing call is a familiar sound in cities because it nests on cathedrals, from Edinburgh (St Mary's) to Canterbury, encouraged by the availability of its favourite prey, pigeons. In the hierarchy of falconry, originally a means of obtaining food and probably originating in Mesopotamia, the peregrine was fit for a prince. *The Boke of St Albans*, Juliana Berners' 1486 treatise on heraldry and hunting, is probably the first known printed book in English by a woman.

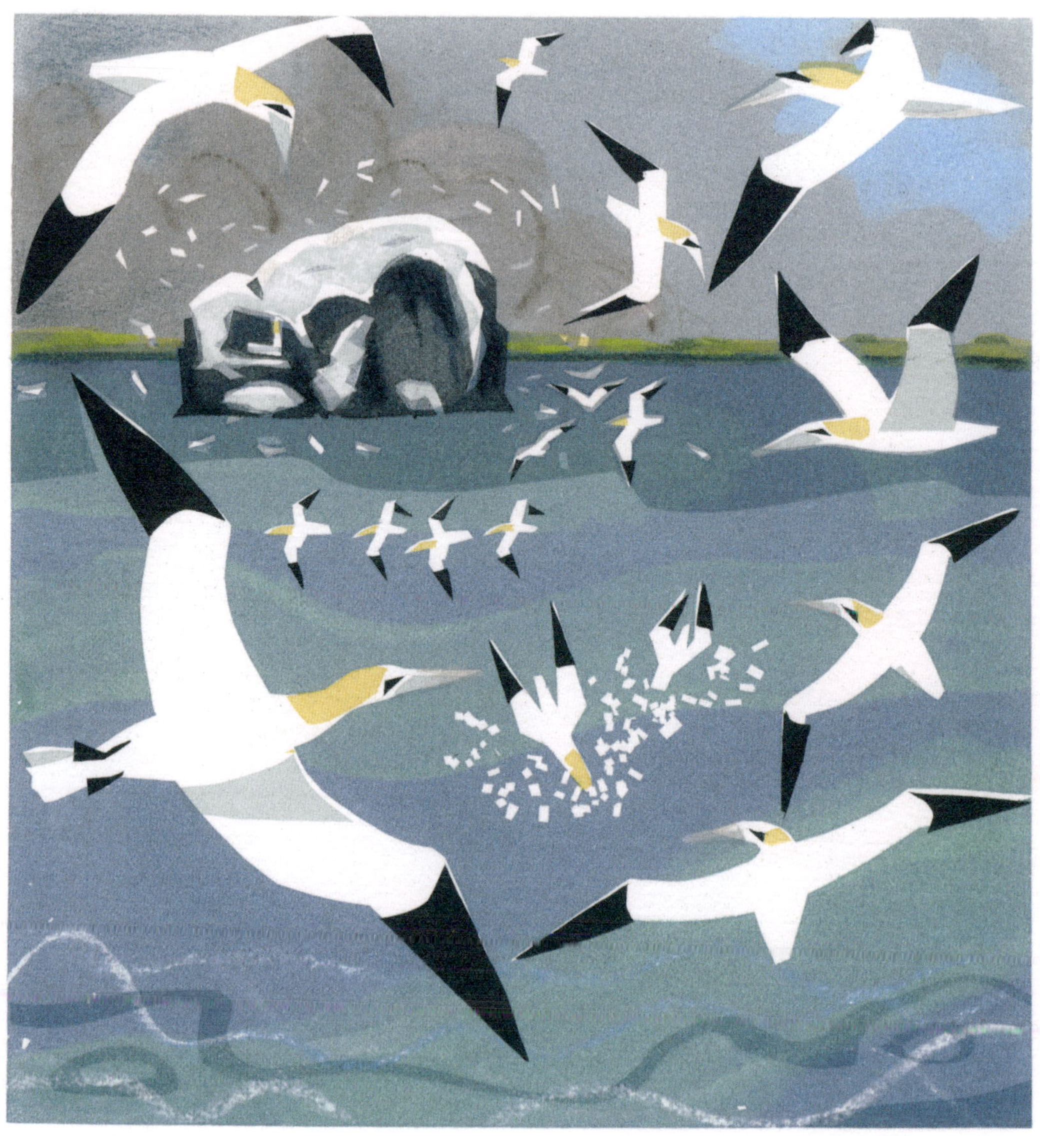

No sight of a bird is more thrilling than the white flash of a fishing **gannet** (*Morus bassanus*) plummeting out to sea at speeds of up to 97km/h (60mph). It can endure the impact due to complex safeguards, from subcutaneous air sacs to nostrils that open inside the bill. The British Isles are home to two-thirds of the world's gannet population (1.8m). The most famous gannetries are mostly on islands; HQ Bass Rock, a mile offshore in the Firth of Forth. Almost half the 75,000-strong colony died from avian flu in 2022, but some survived infection – indicated by a colour change in their irises from pale blue to black – which encourages optimism.

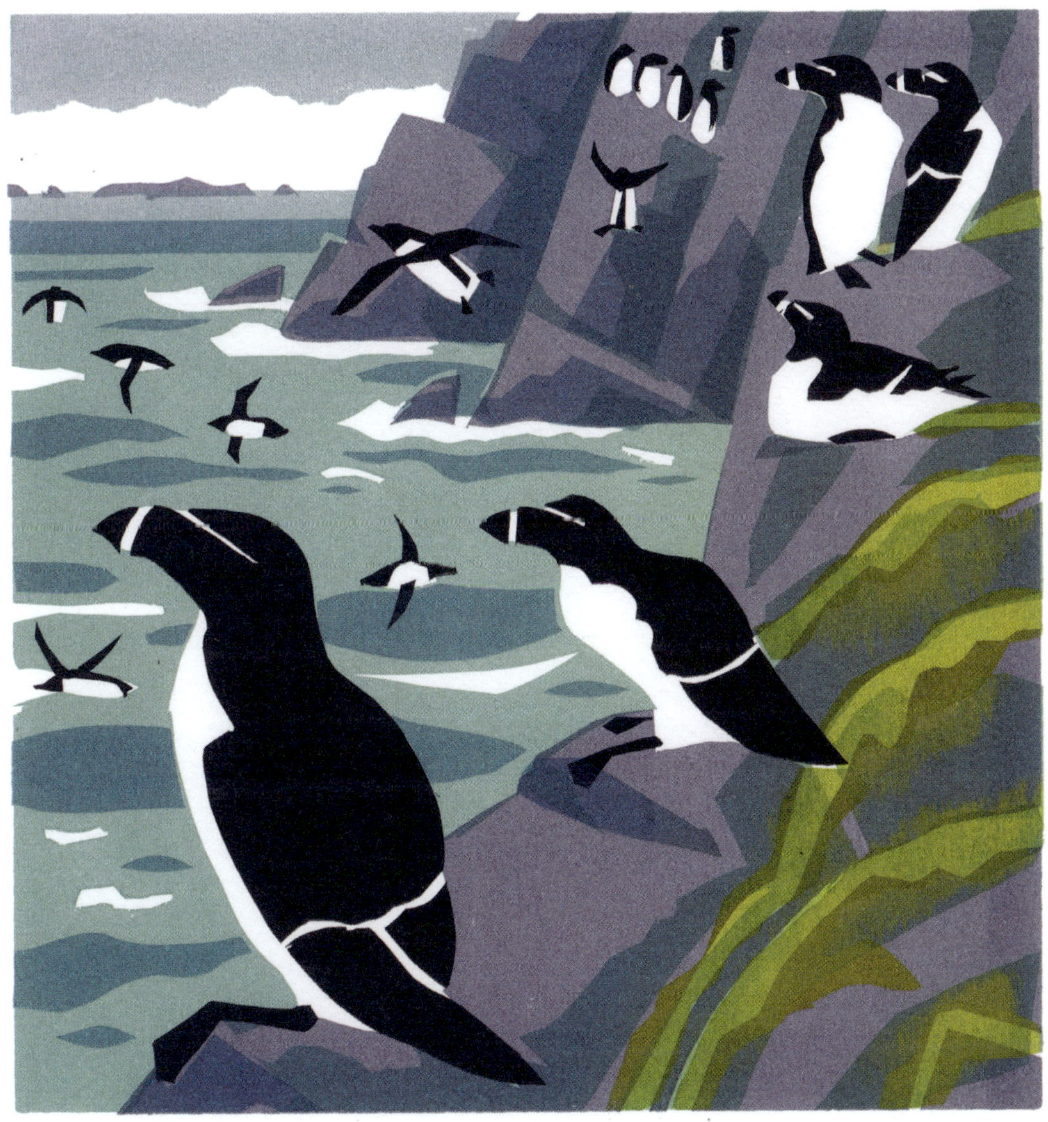

The **razorbill** (*Alca torda*), like the guillemot and puffin, is of the auk family – a smaller version of the fatally flightless great auk, extinct in 1852. In French, it is *petit pengouin* from its upright gait. The single, speckled egg is always clean, unlike the more spectacular but mired guillemot's, because the razorbill nests alone in a crevice, not hugger-mugger. In 1939, at royal request, ornithologist Ronald Lockley (1903–2000) presented a razorbill to the young Princesses Elizabeth and Margaret. It bit him during the handover. The event was one of the UK's first live TV broadcasts.

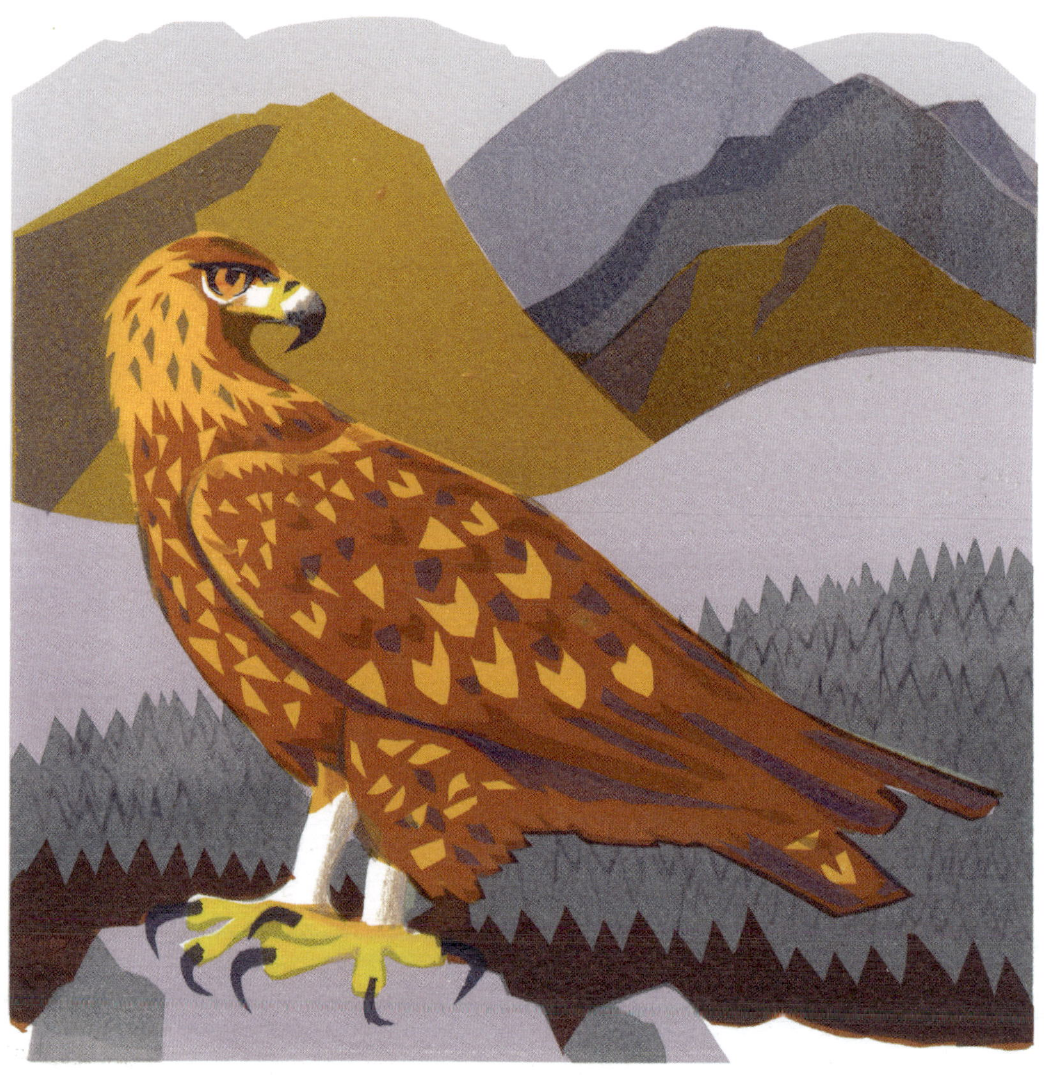

The wrinkled sea beneath him crawls;
He watches from his mountain walls,
And like a thunderbolt he falls.

Alfred, Lord Tennyson (1809–92), from 'The Eagle'

In her memoir (*The Gamekeeper*) Portia Simpson describes a catnap on the Isle of Rum. She awoke to a rush of air. A **golden eagle** (*Aquila chrysaetos*), talons outstretched, was almost upon her. Eagles scavenge but can also kill an insignificant meadow pipit. Buzzards are called 'tourist eagles', as wishful bird lovers often mistake the two: there are 63,000 ubiquitous buzzard pairs but only 440 golden eagle pairs, almost all in Scotland's Highlands and Islands.

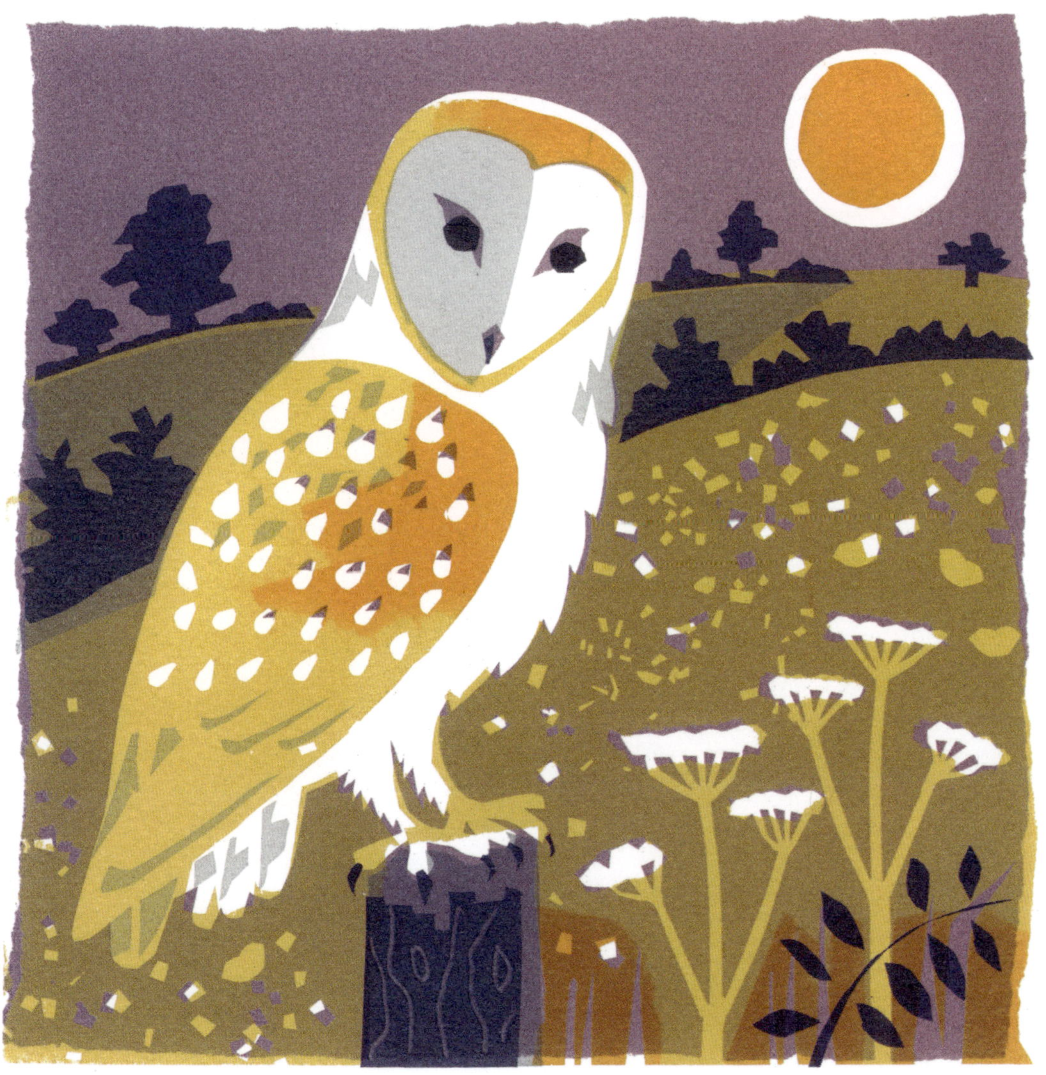

In Berwickshire a forester friend gave me the abandoned owlet of a **barn owl** (*Tyto alba*, 'white') to save. My mother's journal described its 'immensely sad, ancient appearance'. She named it Melchior. Feathered, it purred when stroked. 13 July: fully fledged, at dusk it was lodged outside. Garden filled with barn owls, making 'snoring and escaping steam noises'. 1 August: Melchior still came to be fed or stroked, tapping for entry, 'as disposed for company as ever'. 20 September: 'Melchior has not been back for nearly a fortnight which is very sad.' 29 September: News of dead barn owl on main road a mile away.

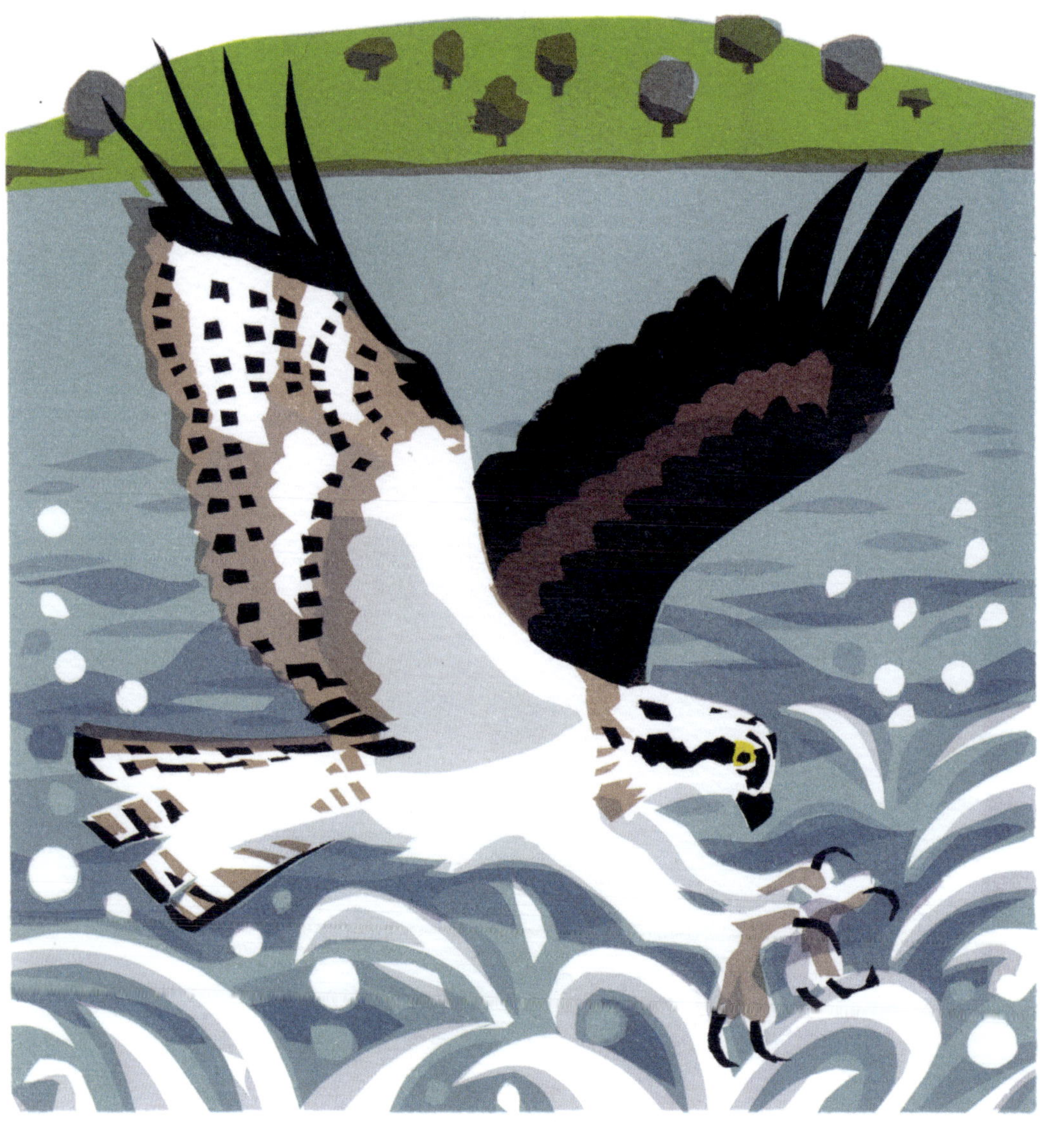

Rarity demanded secrecy when, in 1955, an **osprey** (*Pandion haliaetus*) pair, migrants from West Africa, nested at Loch Garten in Scotland, the first UK nesting since 1916. Colonisation, aided by chick relocation, security measures and the birds' willingness to inhabit artificial nests, followed. By its 60th anniversary in 2019, Loch Garten Nature Centre had received 2.75m human visitors. There are now 250 UK breeding pairs, mostly in Scotland but also Wales and England, as far south as Poole Harbour, Dorset. In 2022, Sacha Dench, founder of Conservation Without Borders, led the 'Flight of the Osprey' team to track three satellite-tagged Scottish juveniles: one was lost at sea, one killed by a goshawk and one survived, having hitched two lifts on ships. A male known as 4K (renamed 'Dobire'), born at Rutland Water in 2013, covered the 5,279km (3,280-mile) journey to Dobire, Guinea, in 26 days.

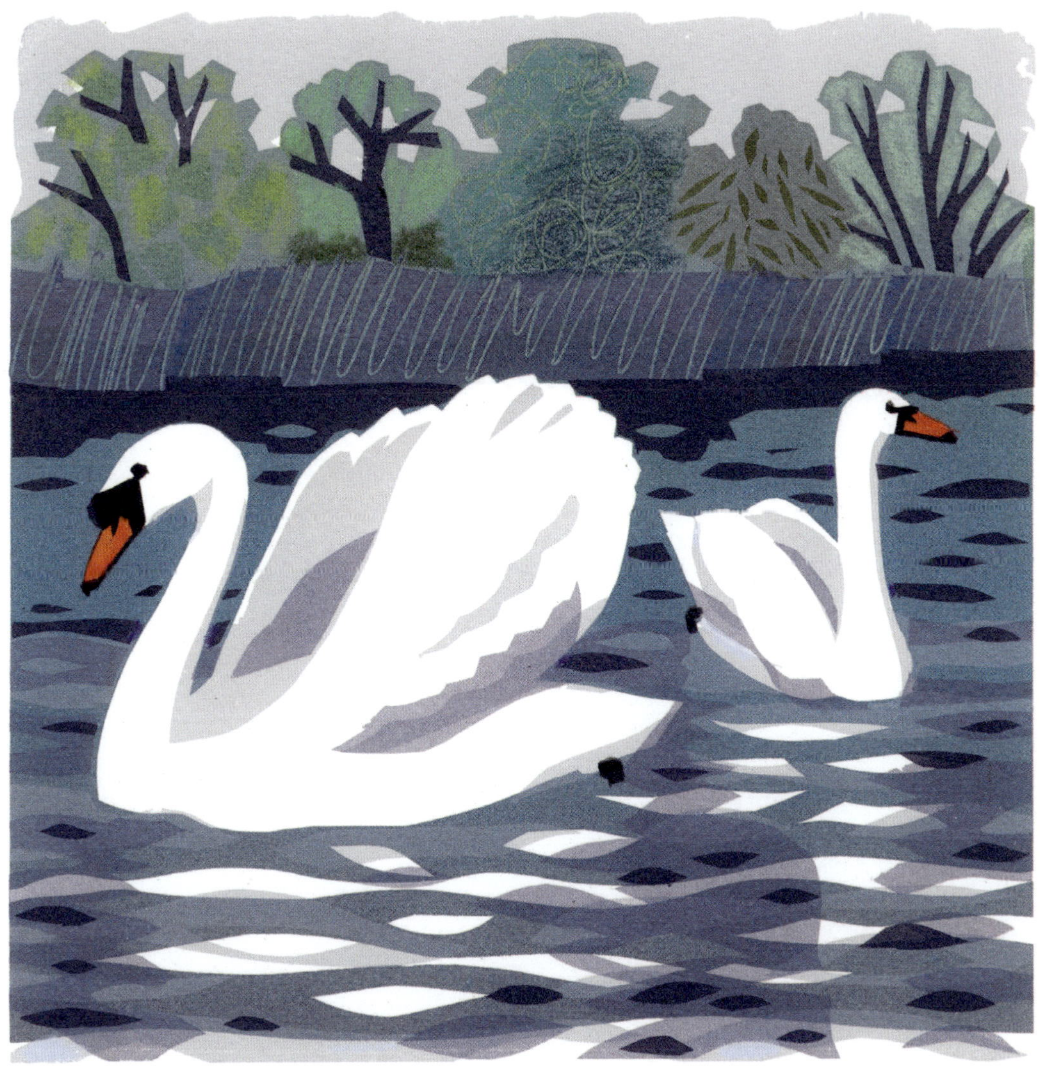

The lyric of 'The Silver Swan' by composer Orlando Gibbons (1580–1625) opens:

> *The silver swan, who living had no note*
> *When death approached, unlocked her silver throat*

Composer John Rutter has tried to encourage a **mute swan** (*Cygnus olor*) to prove it could at least hint at its 'swan song'. So far its best effort is a snort. Traditionally, unmarked swans are Crown property, and the sovereign is Seigneur of the Swans. Thames mute swans beyond Sudbury are counted at the annual count or Swan Upping ceremony. William Shakespeare (1564–1616), born in Stratford upon Avon, was described in his friend Ben Jonson's memorial poem as 'Sweet Swan of Avon'.

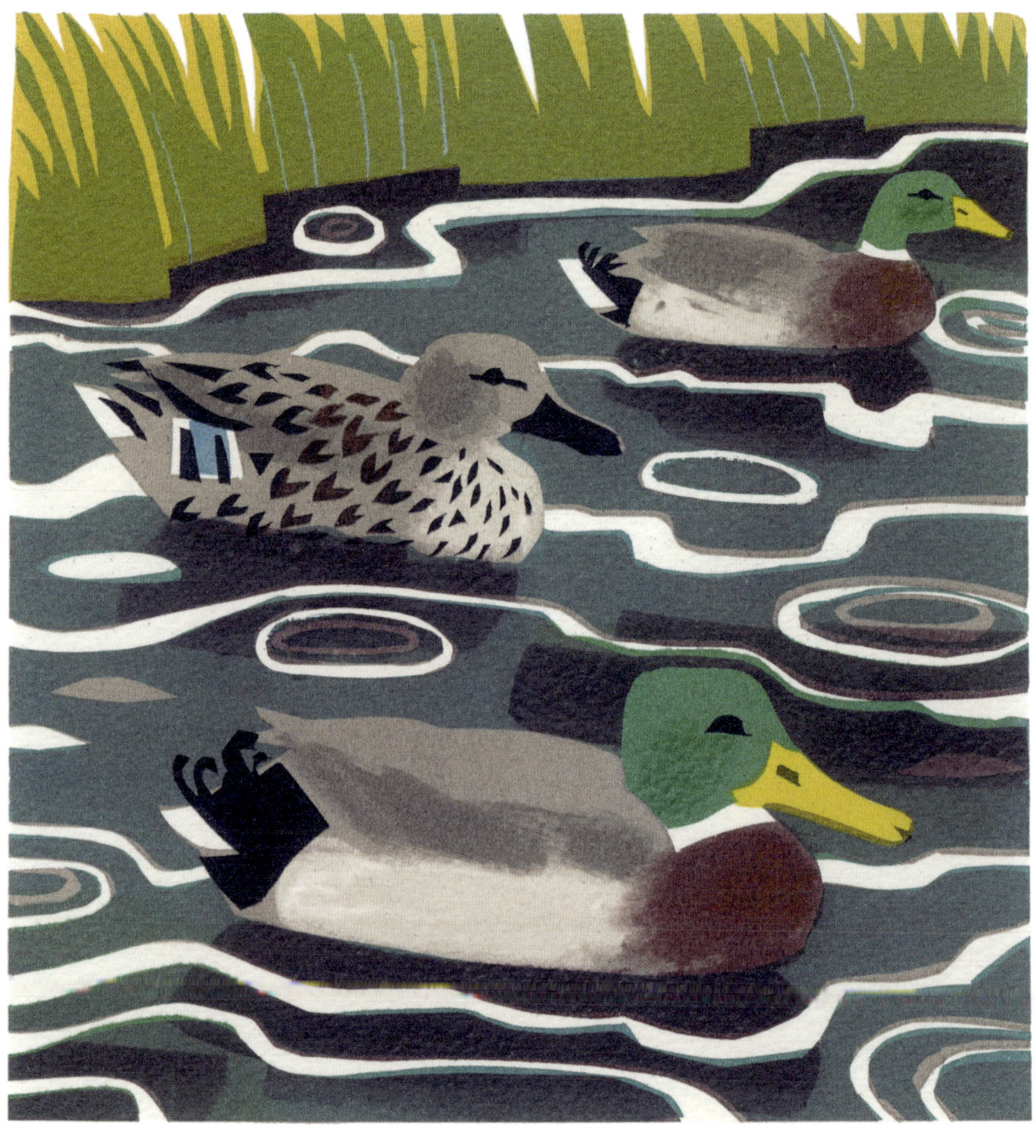

Of all ducks, the female **mallard** (*Anas platyrhynchos*) is the one that quacks the most. The mallard's semi-domestication has generated 20 hybrid 'farmyard' species. It can nest as early as February, sometimes high in trees and tall buildings. Ducklings float from their nests to the ground and are escorted by the mother to the nearest water. Shy in the wild, mallard will happily share a busy pavement in town. Seen on the wing at 6,400m (4 miles), it is the second highest-flying duck after the ruddy shelduck. *Mallard* (fastest-ever steam engine, 202 km/h or 126mph, 1938) was designed by bird-loving engineer Sir Nigel Gresley (1876–1941). Other of his engines were *Bittern*, *Gannet*, *Golden Plover*, *Kingfisher* and *Flying Scotsman* (first to reach 161 km/h or 100mph). It and *Mallard* are at the National Railway Museum, York.

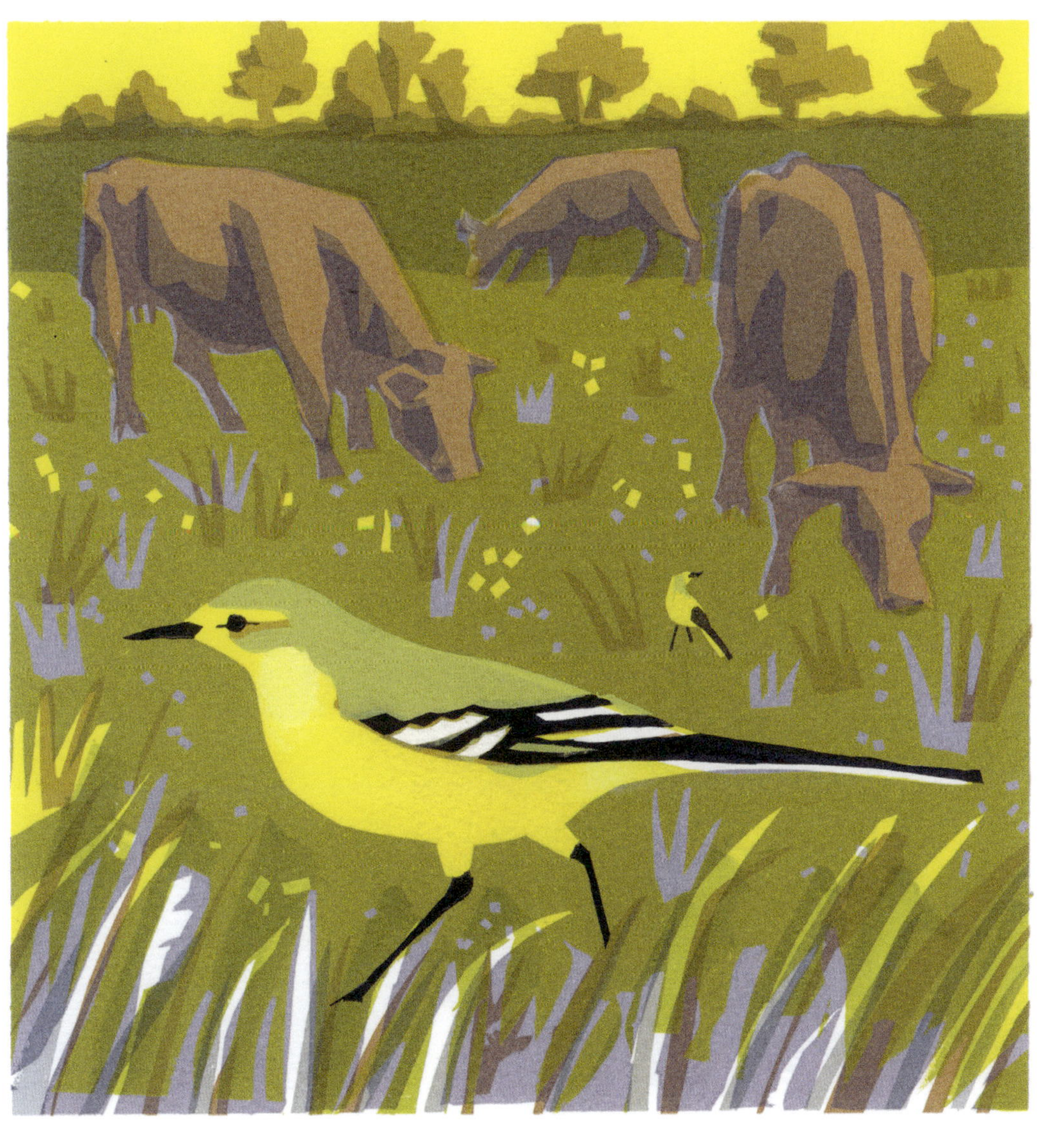

One folklore name for the dazzling **yellow wagtail** (*Motacilla flava flavissima*) is 'sunshine bird', *flavissima*, 'yellowest of all'. England is its principal migratory choice, with Europe host to 10 less-vivid subspecies. In England, water meadows and marshland are its favourite habitat; their draining the principal reason for its decline. In Scotland, Wales and Ireland it is exceptionally rare. A helpful adaptation is its willingness to breed in potato fields – the broad leaves shelter its nest on the ridges, while the tunnels offer hidden escape.

Such safety-places little birds will find

John Clare, from 'Yellow Wagtails Nest'

Families join forces to form flocks for the flight to Africa at summer's end.

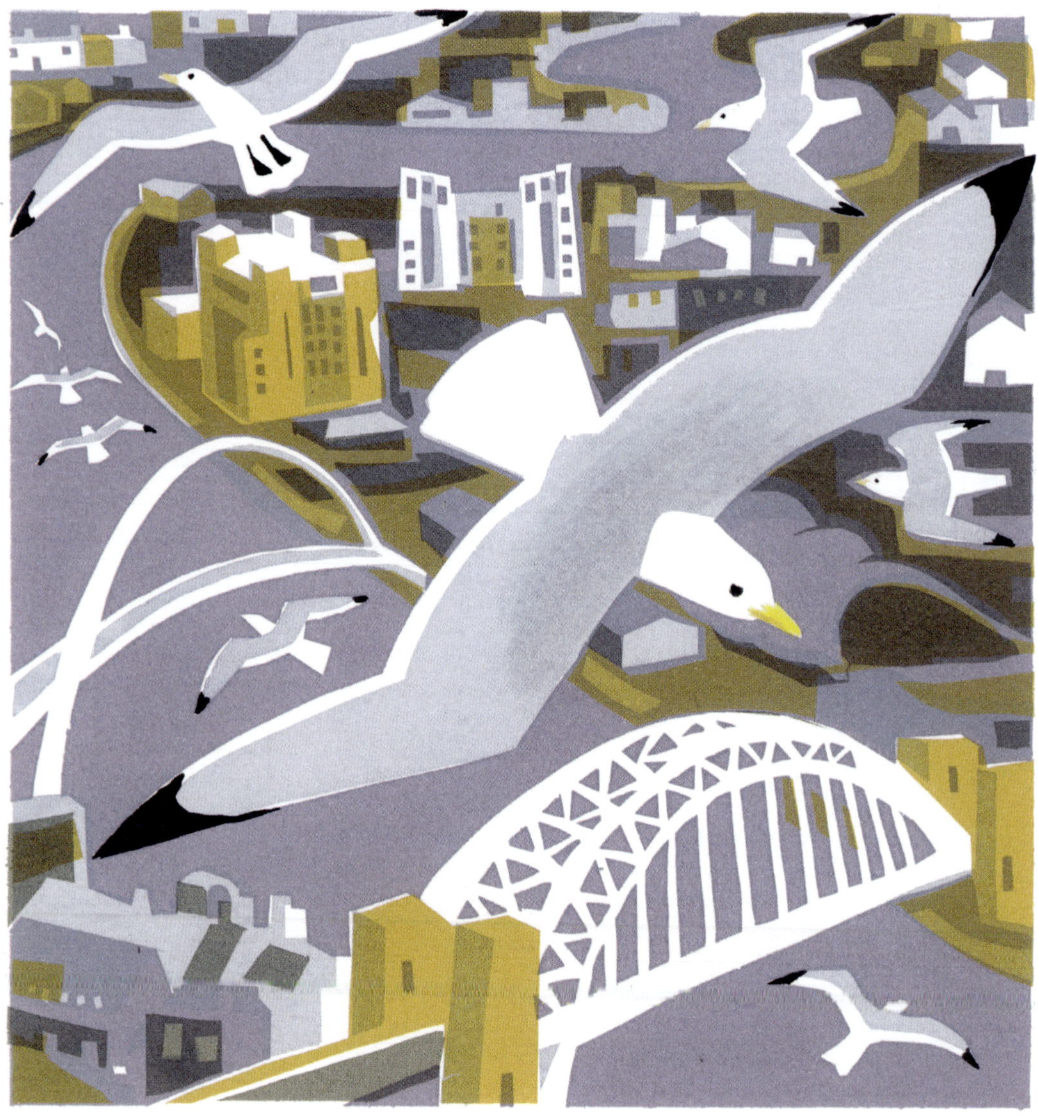

If any gull deserves the name 'seagull', it is the **kittiwake** (*Rissa tridactyla*), which only comes ashore to breed. Yet the most famous kittiwake colonies are in Newcastle upon Tyne – especially on the ledged façade of the Baltic Centre for Contemporary Art on the Gateshead bank of the Tyne, near the 'blinking eye' Millennium Bridge. When the Baltic Flour Mill was converted into the Baltic Centre of Contemporary Art the unwelcome kittiwakes were provided with a specially built artificial tower nearby. Many refused to be re-housed and the Centre's resistance soon turned into grateful acceptance. A viewing platform was built, allowing visitors to come within a few feet of 201 nesting pairs. 'Kittiwake Cam' completes the accessibility of this most urban of seabird colonies.

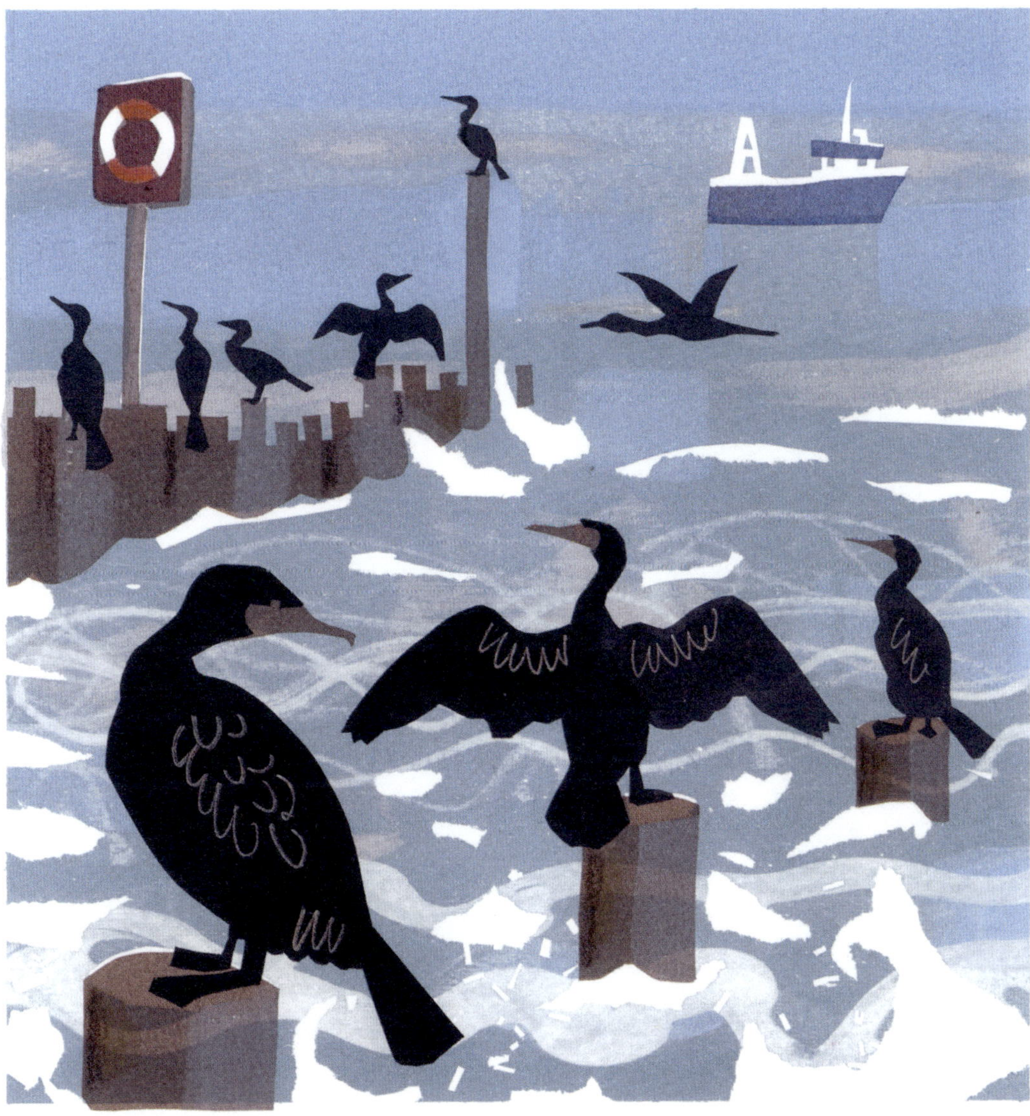

That birds descend from dinosaurs is easy to imagine of the **cormorant** (*Phalacrocorax carbo*), with its reptilian neck and head, plumage demarcated like scales, stiff-quilled tail and upright carriage. Fossils show that the earliest classified bird *Gansus yumenensis*, originating 105–110m years ago, was similarly built. In the Far East, cormorants are used for fishing, the neck ringed to stop them swallowing the catch. Painter Henri de Toulouse-Lautrec (1864–1901) walked his pet cormorant on a lead along seaside promenades to scare old ladies. He also introduced it to absinthe. 'It takes after me, it's developed a taste for the stuff,' he said.

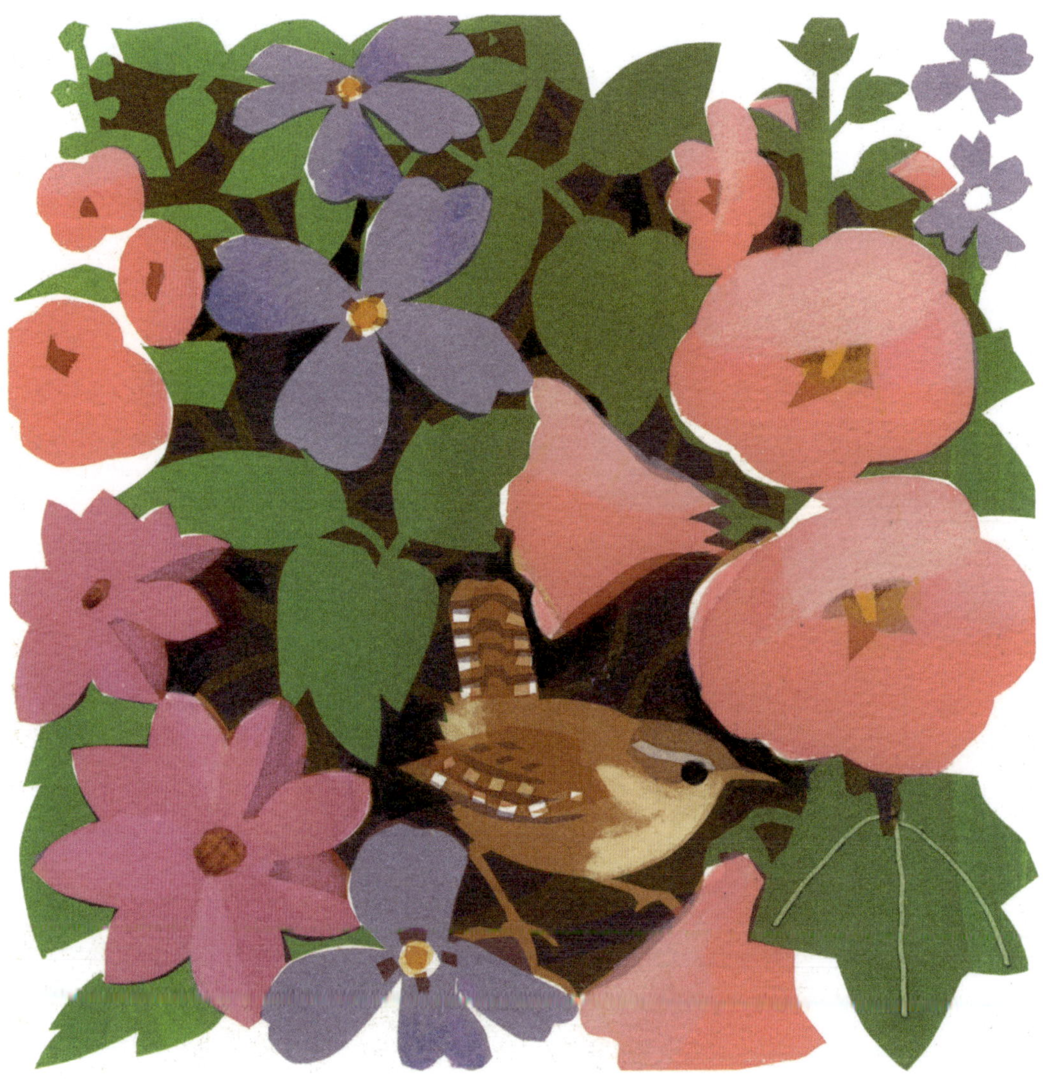

The **wren** (*Troglodytes troglodytes*) is as ubiquitous as the spiders it eats, found from city centres to islands so remote it has evolved insular sub-species. Eleven million UK territories make it the country's most populous bird. Its torrid, repeated song – per unit weight 10 times louder than a cockerel's crow – belies its mouse-like size. The male offers the female several domed nests; the one she chooses is then lined with feathers. For winter warmth, many wrens crowd into a single nest – 60 in one nest box the record.

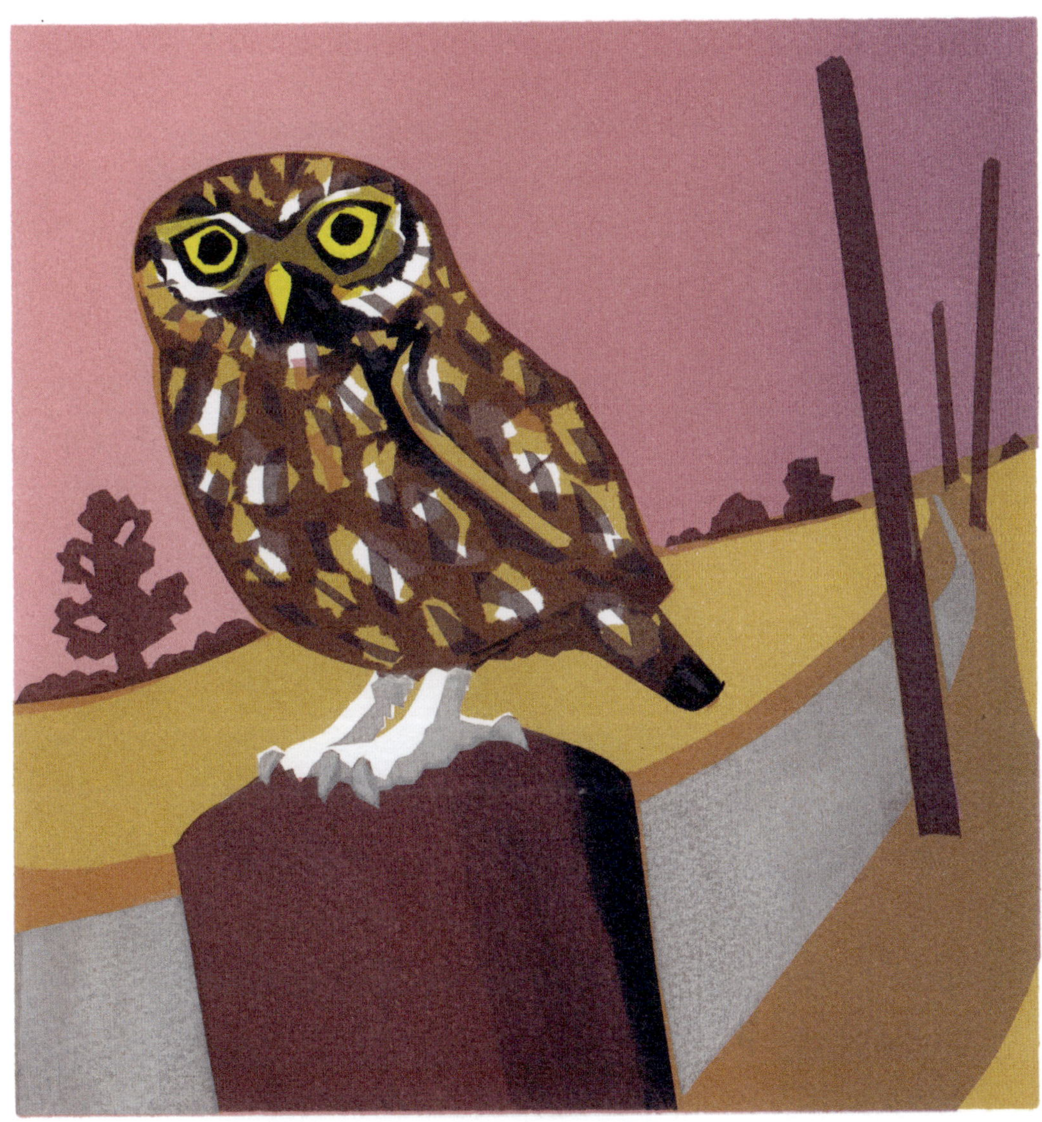

Athena, daughter of Zeus, mightiest of gods, and Metis, the wisest, had, as her sacred bird, the **little owl** (*Athene noctua*), hence 'wise owl'. The Greek €1 coin bears its image, copied from a 500 BC coin. Florence Nightingale (1820–1910), nurse and pioneer statistician, had a pet little owl called Athena, now stuffed and on display in London's Florence Nightingale Museum in St Thomas's Hospital. Pablo Picasso (1881–1973) had Ubu, who appears in various forms of his art. If Ubu snorted disapproval, Picasso swore back. The little owl was introduced to the UK in the nineteenth century, notably at Lilford Hall, Northamptonshire, by the 4th Lord Lilford (1833–96). In 2023, one was killed by a red kite – at Lilford.

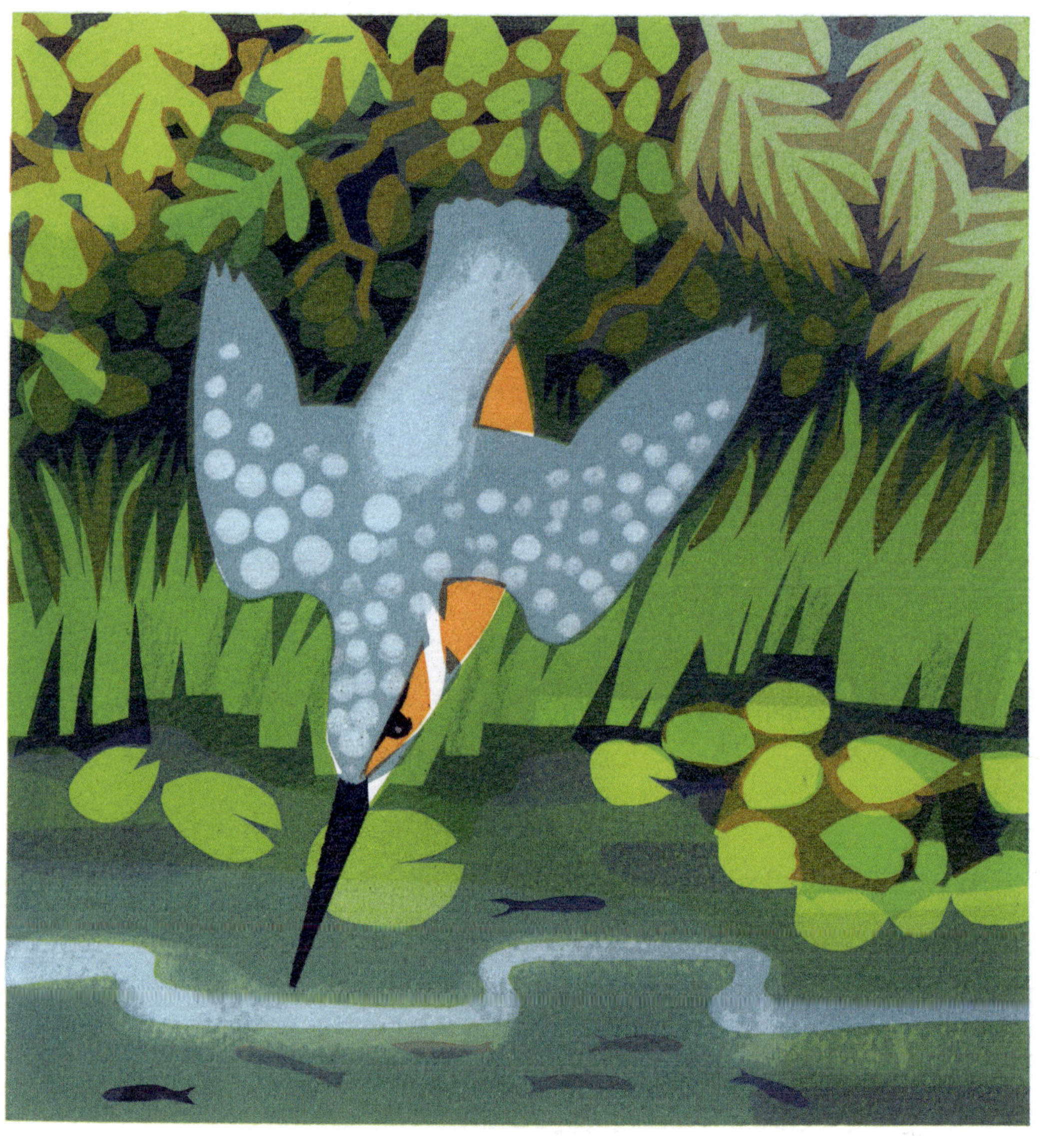

The viscous air wheresoe'er she fly,
Follows and sucks her azure dye...
And men the silent scene assist,
Charm'd with the sapphire-winged mist.

Andrew Marvell (1621–78), from 'Upon Appleton House'

The **kingfisher** (*Alcedo atthis*, 'king of fishers') is aptly named. A pair must catch more than 100 fish a day to feed a brood of up to seven nestlings. The nest burrow rises vertically at its end so nestlings will be temporarily safe should a flood submerge the entrance. Glacial winters can send kingfishers to the coast for open water.

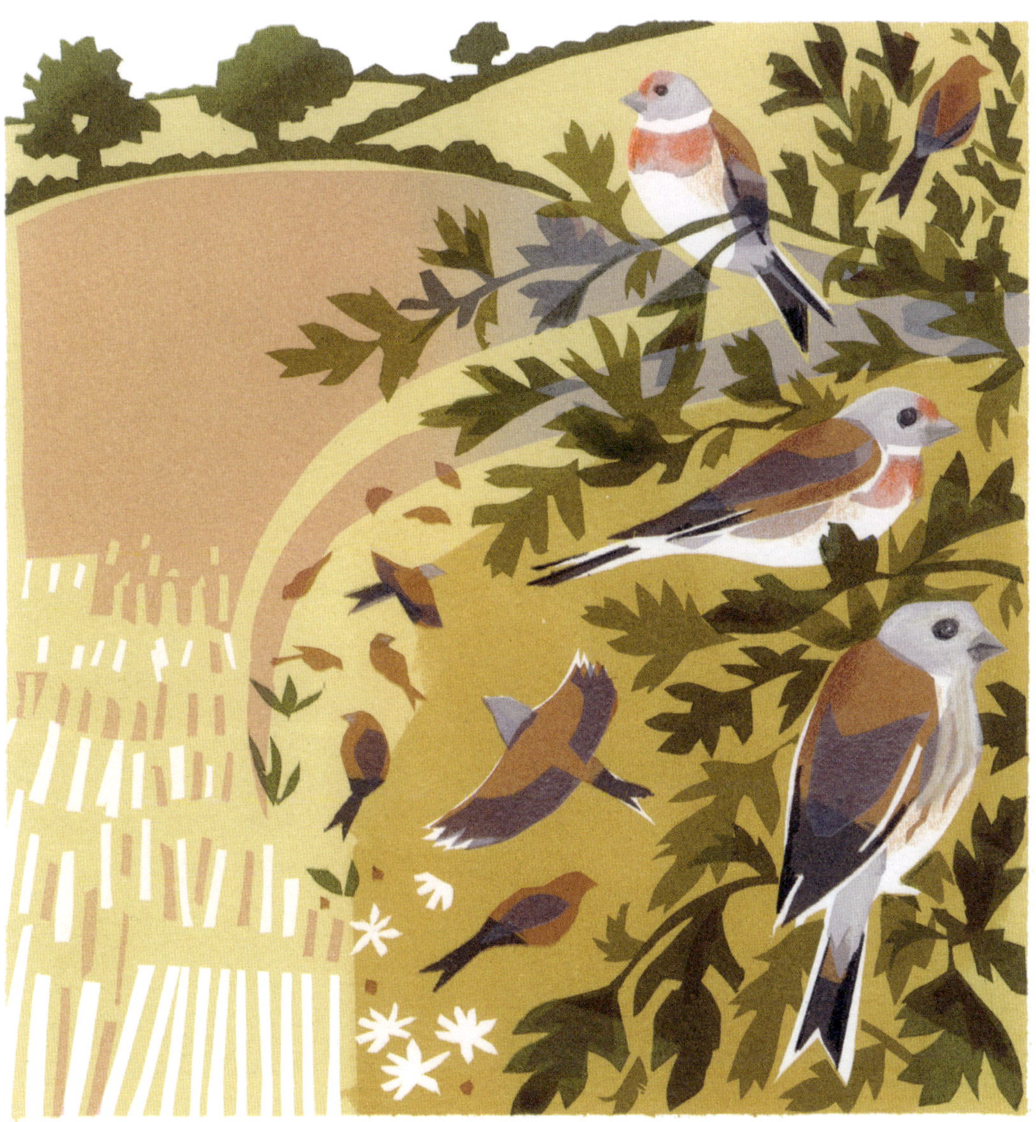

The **linnet** (*Linaria cannabina*) was once a popular cage bird, as composer Charles Collins (1874–1923) and lyricist Fred Leigh (1871–1924) testify in their 1919 song, 'Don't Dilly Dally on the Way,' subtitled 'The Cock Linnet Song': *I walked behind with me old cock linnet*. In the wild, gorse or furze ('whin', Scotland) is its favourite habitat. The UK population (560,000 territories) has fallen 60 per cent since 1970, meaning *'And evening full of the linnets wings'* from 'The Lake Isle of Innisfree' by W. B. Yeats (1865–1939) no longer evokes the once-familiar country sound.

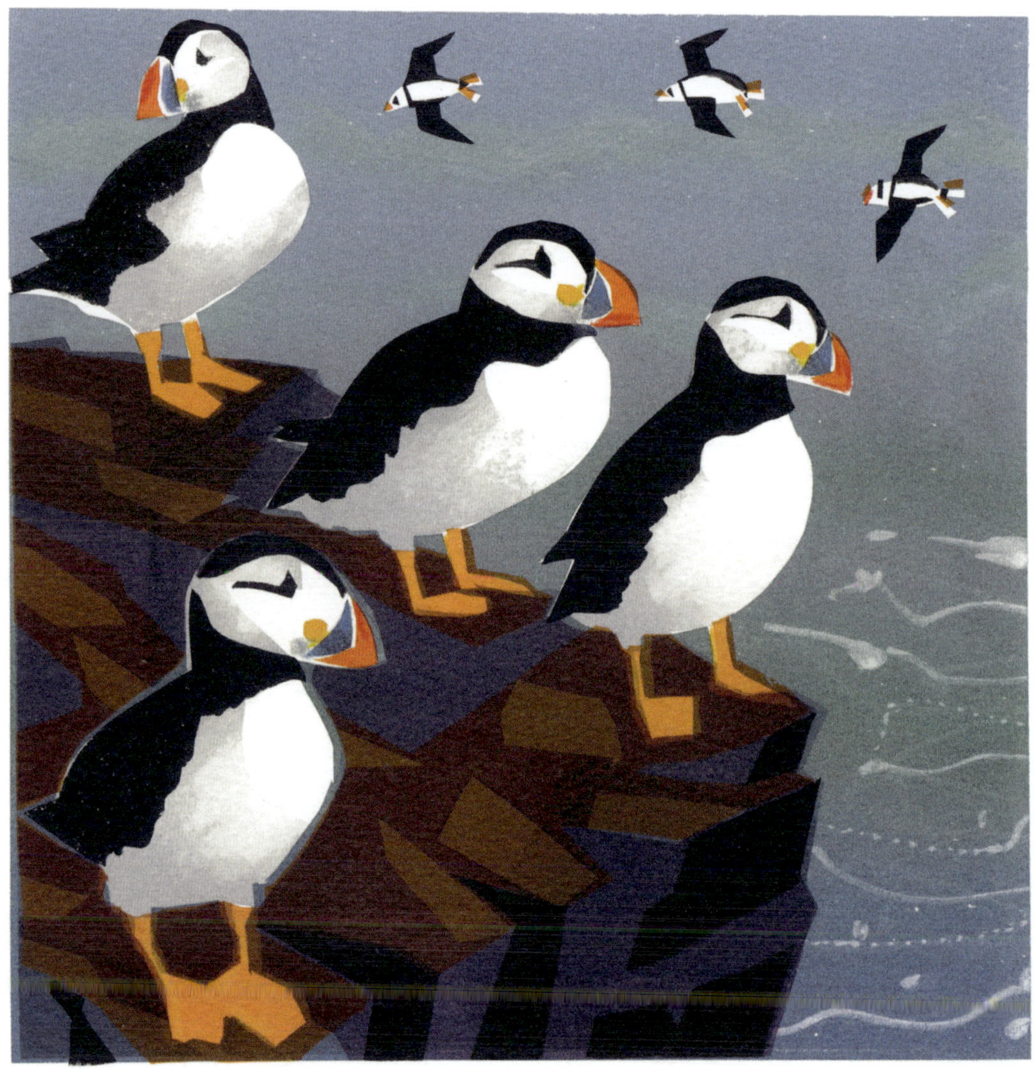

The **puffin** (*Fratercula arctica*) uses its clownish bill to dig its nesting burrow, but it can also catch and clamp numerous sand eels (the record 61), diving to depths of 60m (200ft). In winter, soberly billed, it wanders the sea. The Bristol Channel's Lundy Island (*lunde*, Norse for puffin) has its private postal service, the world's oldest, with puffin stamps and puffinage payment, deliverable within the UK. Lundy is owned by the Landmark Trust, providing 33 properties for holiday lets from castles to cottages. Breeding seabird numbers at Lundy have dramatically increased since the rat extermination programme (2002–06): from 7,351 (2000) to 40,000 (2023).

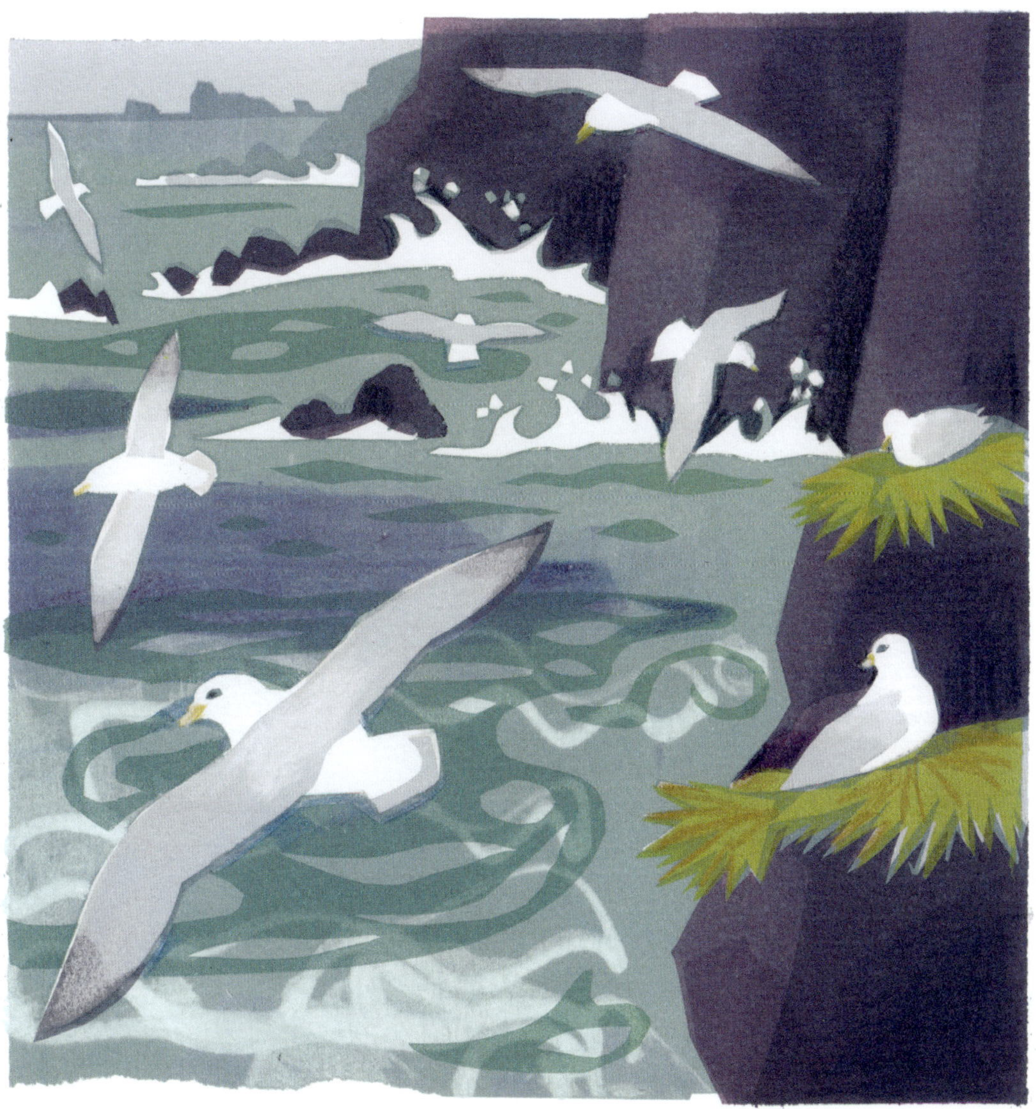

Once confined in the British Isles to St Kilda, the **fulmar** (*Fulmarus glacialis*) in the twenty-first century is ubiquitous on rugged coasts, thanks to its adaptability, combined with surplus fish dumping (230,000 tonnes discarded in EU waters, 2019). As a petrel, it has 'tubenose' nostrils for an enhanced sense of smell. Its elegant, gliding flight contrasts with 'foul' defensive vomiting, hence the 'foul maa' origin of its name. 'The bird had a really good aim,' said Cody Weishaar (aged 7) after becoming the youngest person to climb Orkney's 137m (449ft) Old Man of Hoy sea stack in August 2022. 'It spat on my face, right on my cheek.'

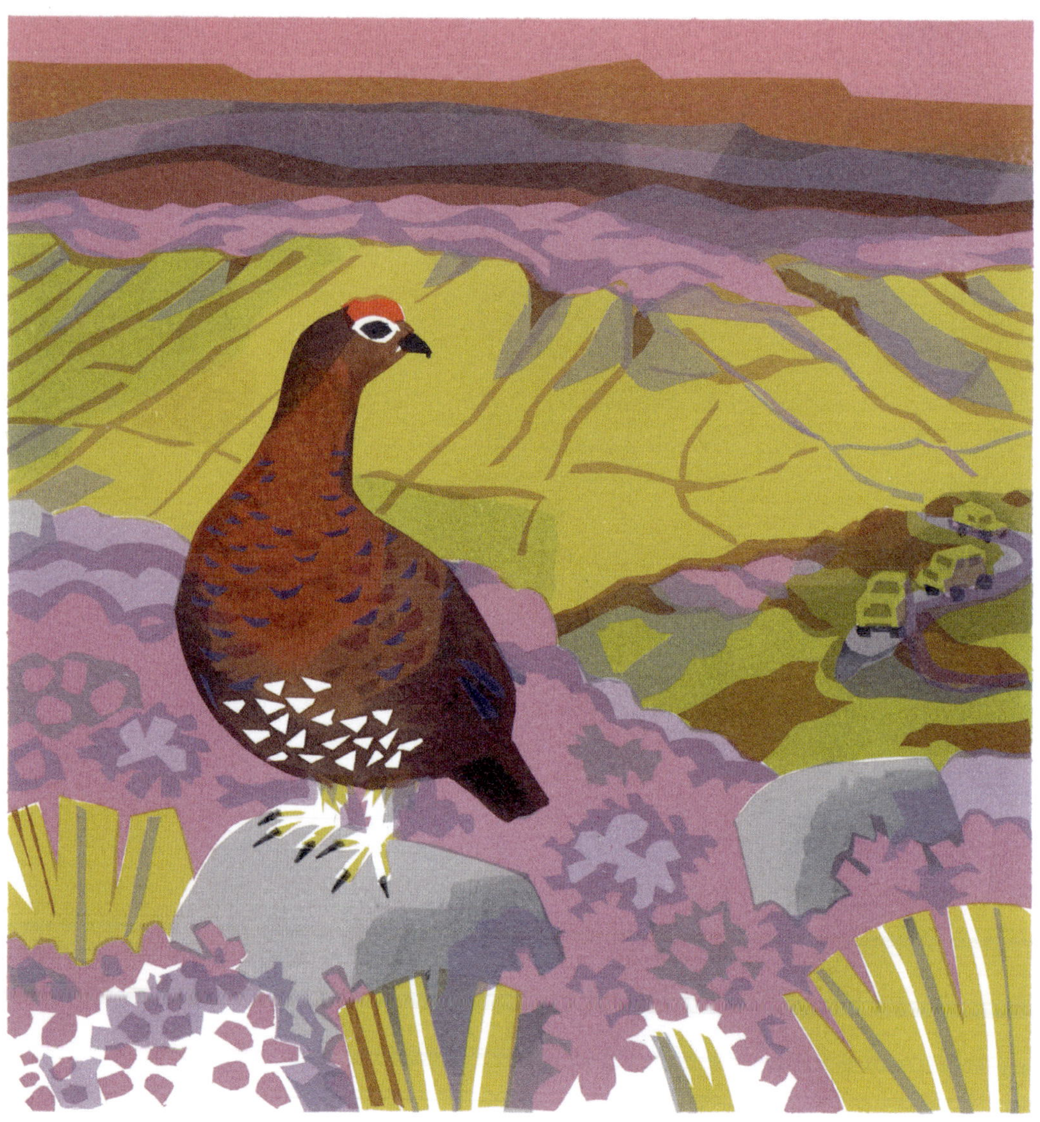

The **red grouse** (*Lagopus lagopus*) was a favourite of W. H. Hudson. Visiting a Highland grouse moor, he wrote in *British Birds* of the bird's 'perfect harmony' with 'that wild and solitary nature amid which it exists'. Unlike pheasants and partridges, grouse are completely wild and cannot be reared. Henry Douglas-Home (*The Birdman*) wrote of a Tomintoul postman who was issued with a car to protect him from a fierce cock grouse. For grouse to survive in numbers, costly protection and habitat maintenance of grouse moors is necessary (Ian Coghill, *Moorland Matters*). In summer, grouse moors are vital breeding havens for many bird species, especially waders.

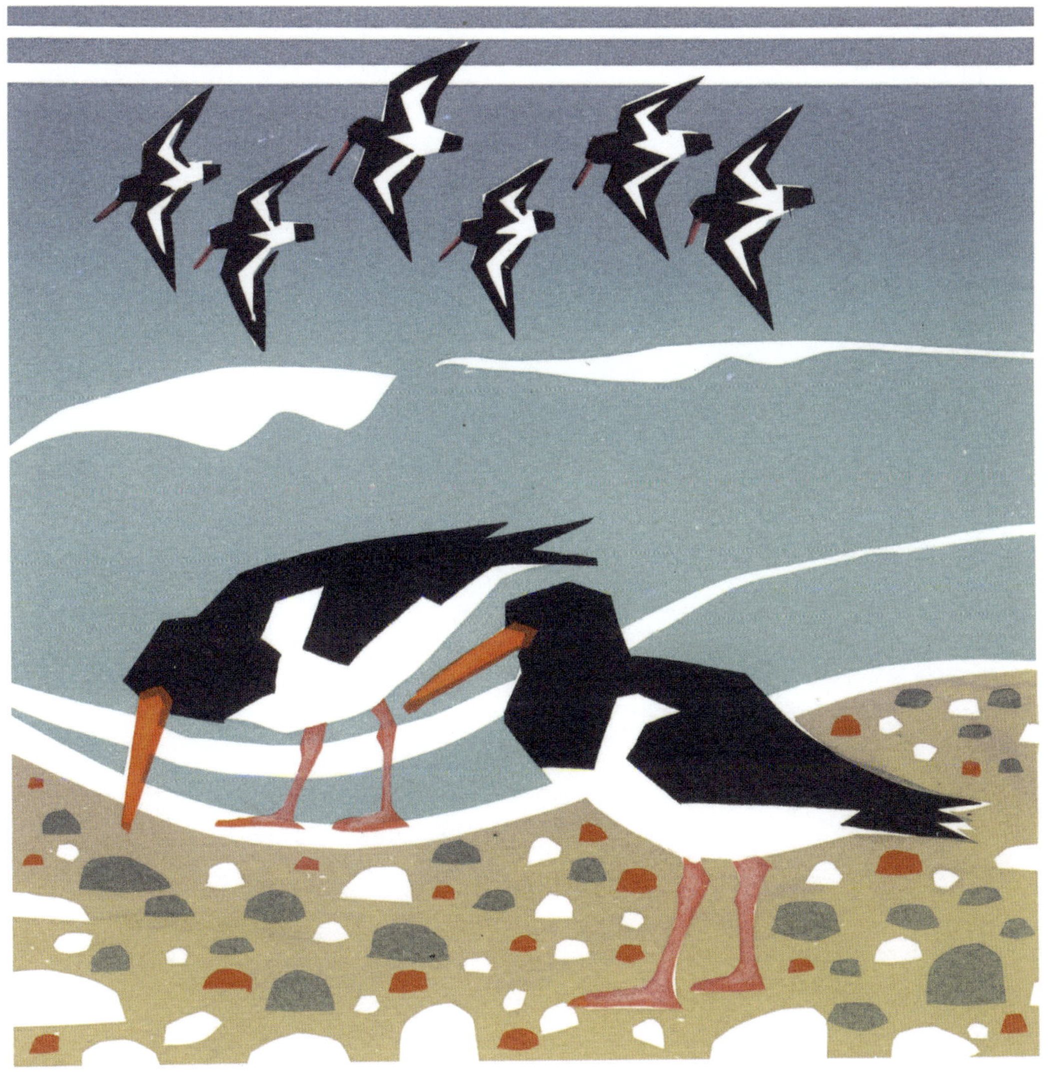

The name of the **oystercatcher** (*Haematopus ostralegus*) is misleading – it prefers mussels, requiring up to 450g of meat daily, equivalent to prising 500 shells. In summer many breed inland, in recent times on town roofs – first witnessed in Aberdeen. Almost half the European population of 300,000 migrants come to the British Isles in winter. A bird aged 35, re-trapped at Wainfleet, Lincolnshire (2002), holds the British longevity record for a wader. Polymath and author Sir Thomas Browne (1605–82) noted the oystercatcher was called a 'dickey-bird' at Wells-next-the-Sea, Norfolk.

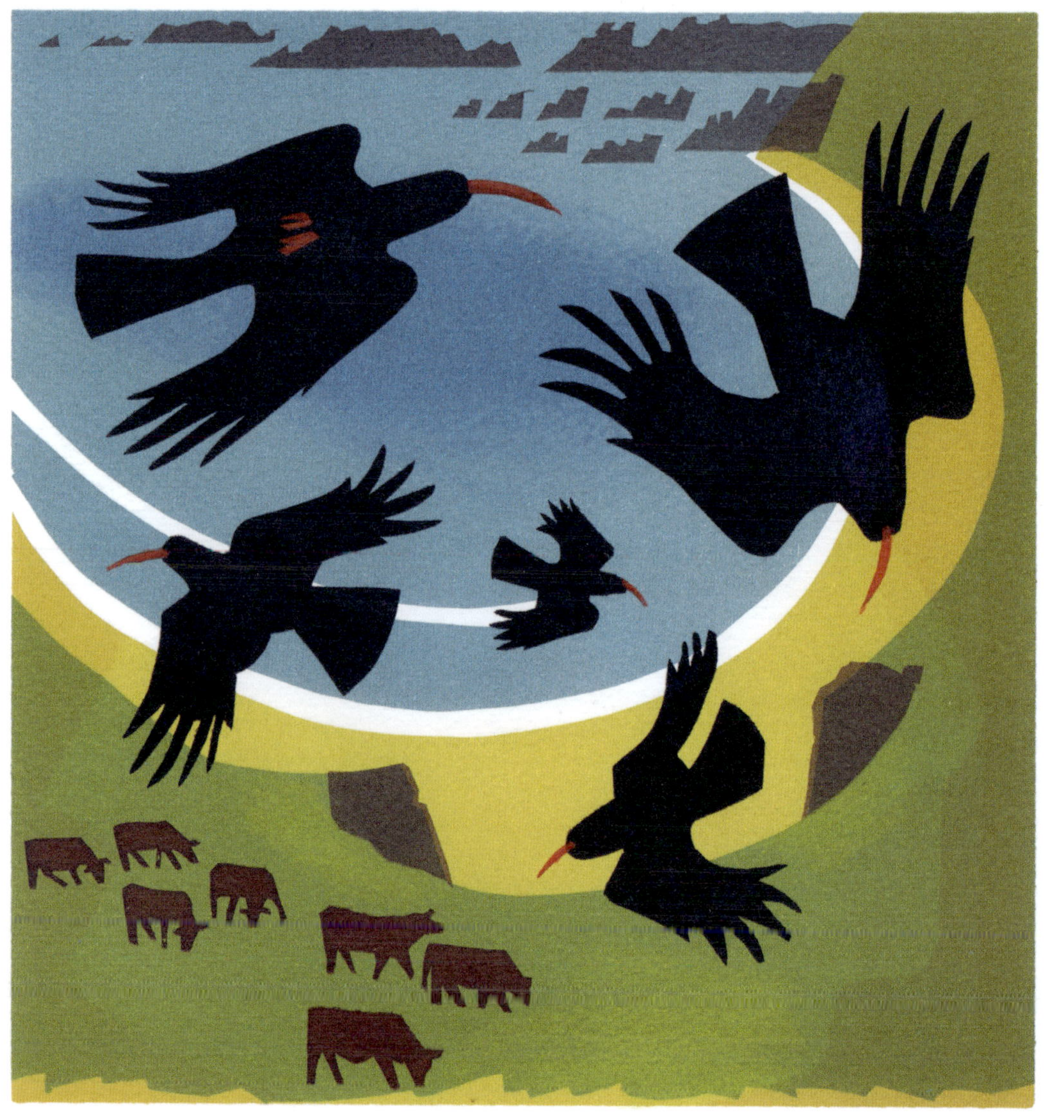

And dizzy 'tis to cast one's eyes so low!
The crows and choughs that wing the midway air
Show scarce so gross as beetles:

William Shakespeare, from *King Lear*, Act iv, Scene vii

Cornwall is the immemorial home of the **chough** or 'Cornish crow' (*Pyrrhocorax pyrrhocorax*). It adorns the coats of arms of Cornwall and Canterbury – its legs and bill blood-red from consoling the assassinated Thomas à Becket (*c.* 1119–70) in the cathedral. Operation Chough, based in Paradise Park, St Ives, sent birds to Kent's Wildwood Trust in 2022 in honour of St Thomas and, it is hoped, to recolonise Shakespeare Cliff (after *King Lear*), Dover.

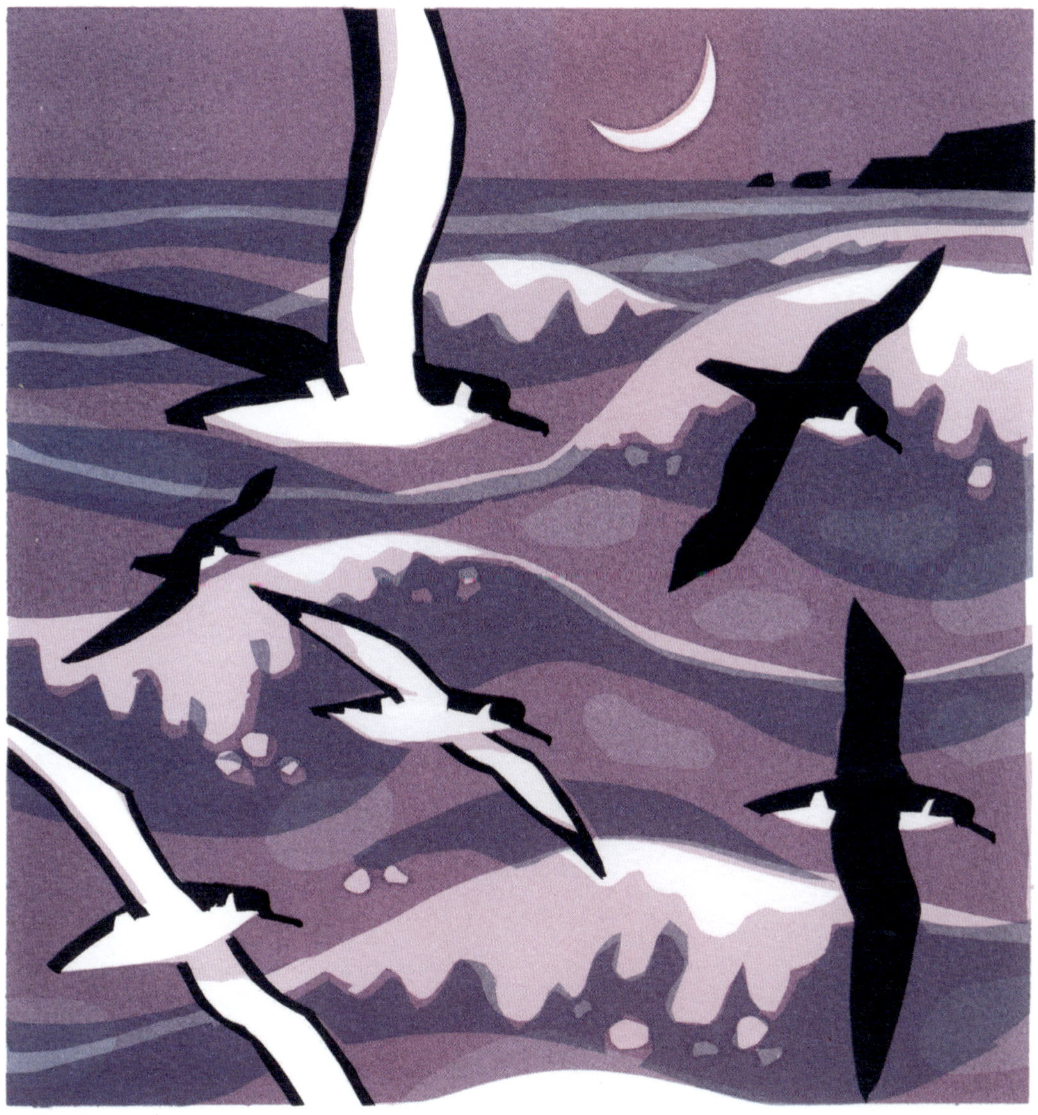

Seventy-five per cent of the global population (410,000 pairs) of the **Manx shearwater** (*Puffinus puffinus*) nest in the British Isles – most on the islands of Skomer, Skokholm, Middleholm (Wales) and Rum (Scotland). In 1926, pioneer conservationist Ronald Lockley leased uninhabited Skokholm for 21 years. Manx shearwaters, summer migrants from the South Atlantic, nest in rabbit burrows. They share parental duties, fishing by day and relocating at night with ghoulish screams to feed the single chick the regurgitated catch. In 30 years, 65,579 Skokholm shearwaters were ringed. Lockley tested their navigation skills to the limit. A bird released in unfamiliar Venice arrived on Skokholm in 14 days. Shearwaters are related to albatrosses, while the puffin (*Fratercula arctica*) is an auk. 'Puffin' comes from the shearwater chick/'puffin', a prized delicacy; puffins still prized fare in Iceland.

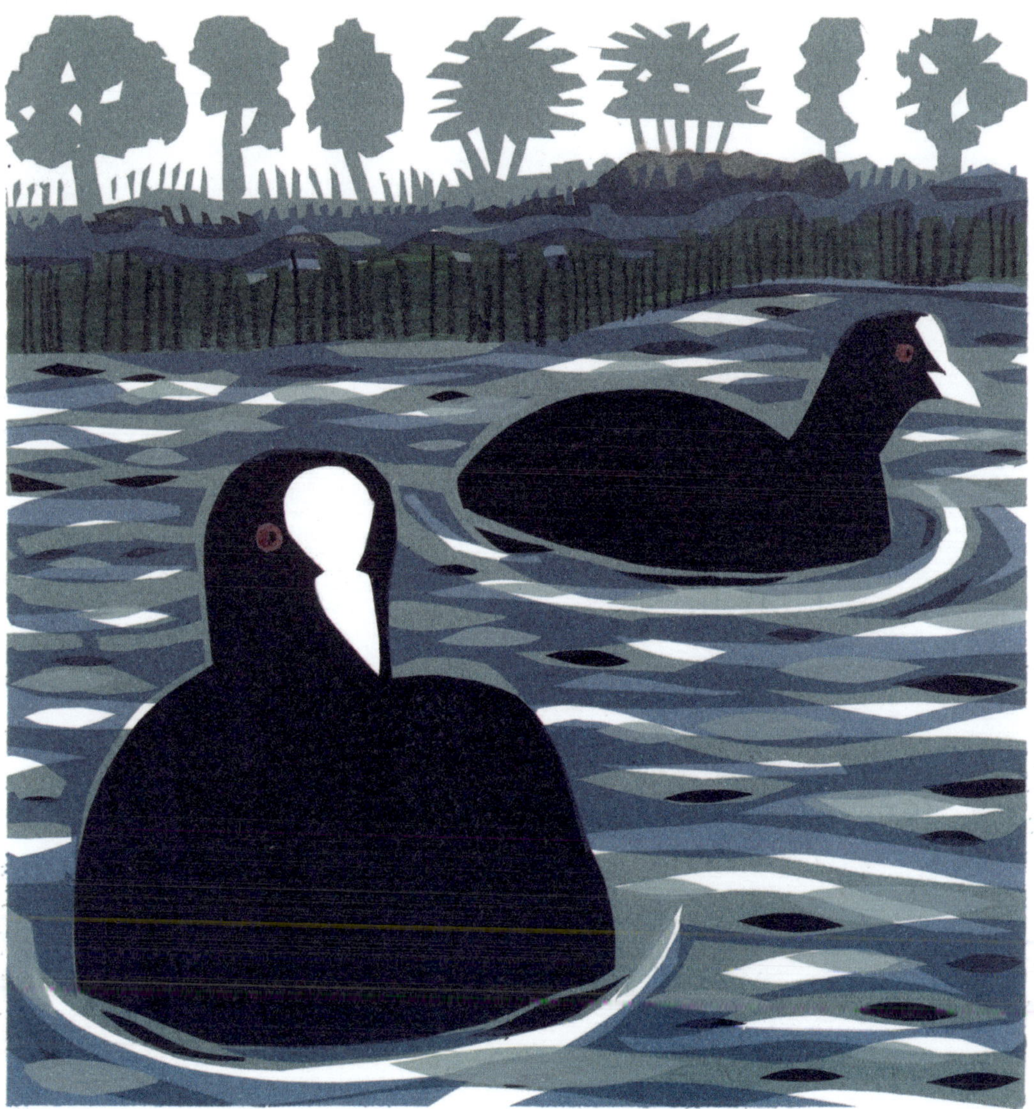

If moorhens' feet seem too large, those of a **coot** (*Fulica atra*) are weird to boot – grey and multi-lobed as if another creature's altogether. Coots are quarrelsome, which reveals the feet as weapons – one bird lying back to flail at its opponent. Unlike moorhens, they are strong flyers – surprisingly, since they always scamper across water to launch an attack or attend a bread-scattering. The phrase 'bald as a coot' derives from their white-skinned foreheads. Coots and moorhens are protected in Northern Ireland but fair game in the shooting season elsewhere in the UK, 1 September - 31 January, though this is little practised. Coot shooting remains a popular field sport in the US (the American coot, *F. americana*, the target).

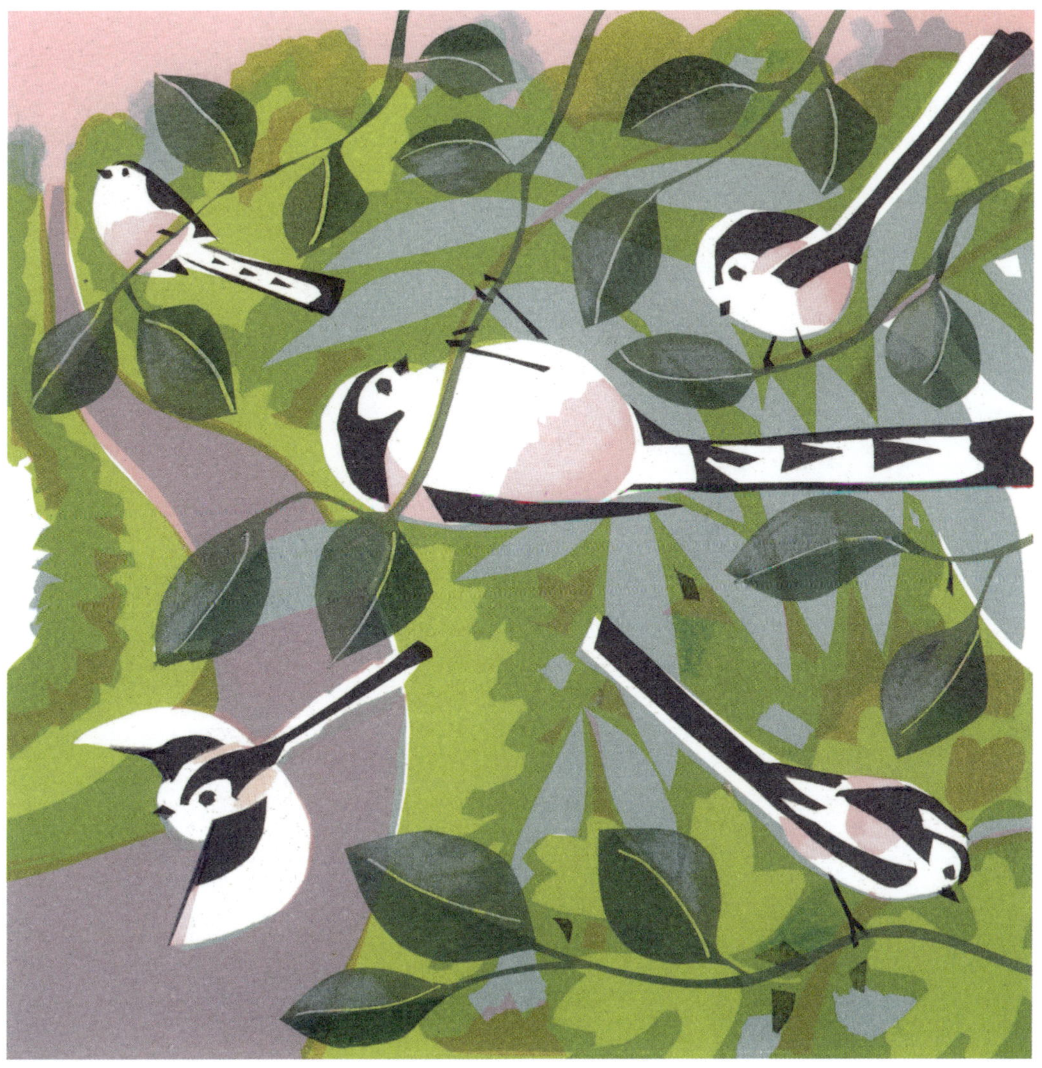

Numbers of the exotic **long-tailed tit** (*Aegithalos caudatus*) have increased by 80 per cent in the UK in 25 years, apparently through adaptation to town life. On warm autumn days, roaming families briefly make fountains of bird baths. The 7.5cm (3in) tail disguises how slight the bird is (weighing no more than a 5p piece). John Clare called it a 'bumbarrel' and described its famous nest of mixed bird feathers (a record 383, *The Birdman*):

Of mosses grey with cobwebs closely tied
And warm and rich as feather-bed within

From 'The Bumbarrels Nest'

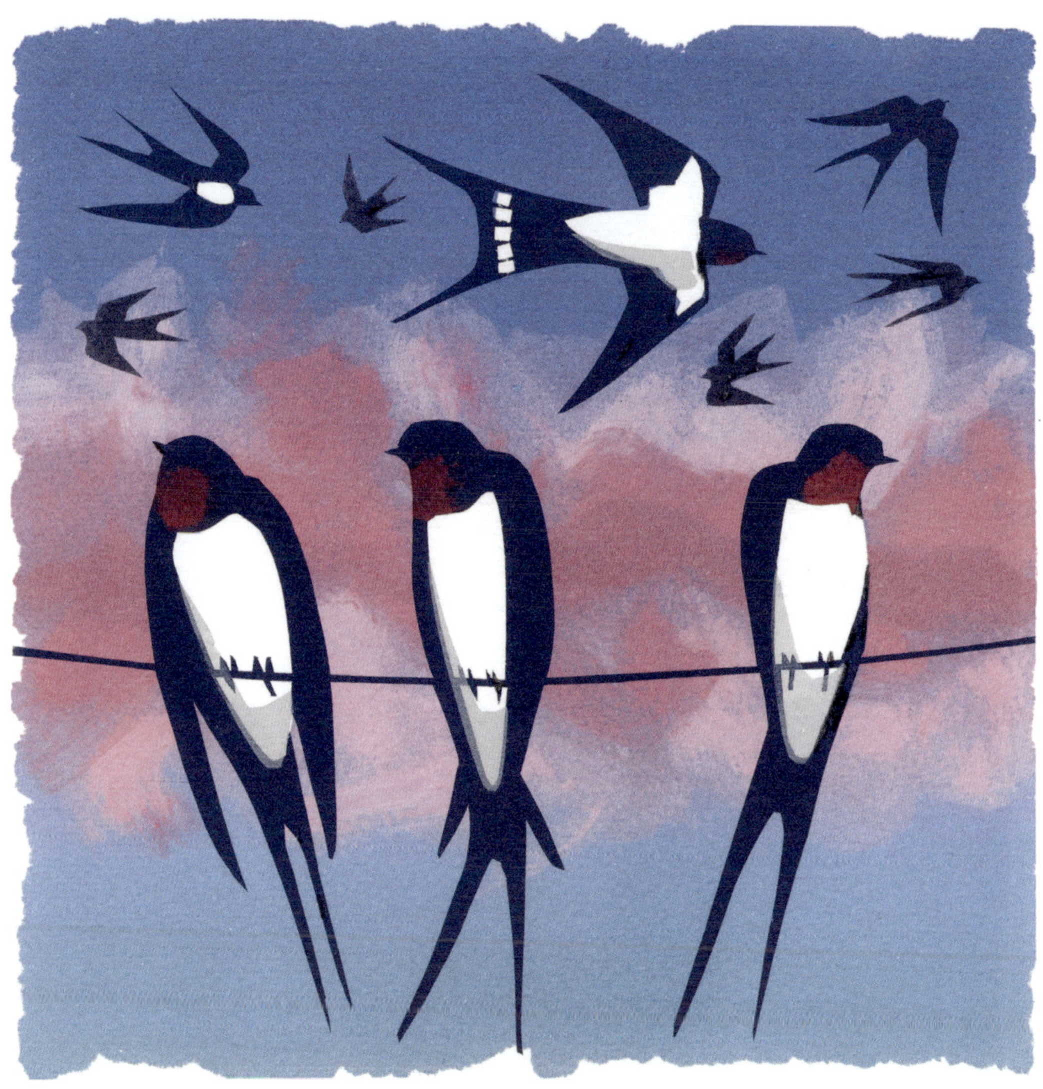

It's dowie in the hint o' hairst,
At the wa-gang of the swallow,
When the winds blaw cauld,
And the burns rin bauld,
And the woods are hingin' yellow.

Hew Ainslie (1792-1878), from 'The Hint o' Hairst'

In autumn, the **swallow** (*Hirundo rustica*) and martins gather in a feeding frenzy at migrating staging posts (*zugunruhe*, no English equivalent), such as the Channel coast above Hastings. Swallows nest in barns. In Spain (Andalusia) they are welcome to breed inside houses.

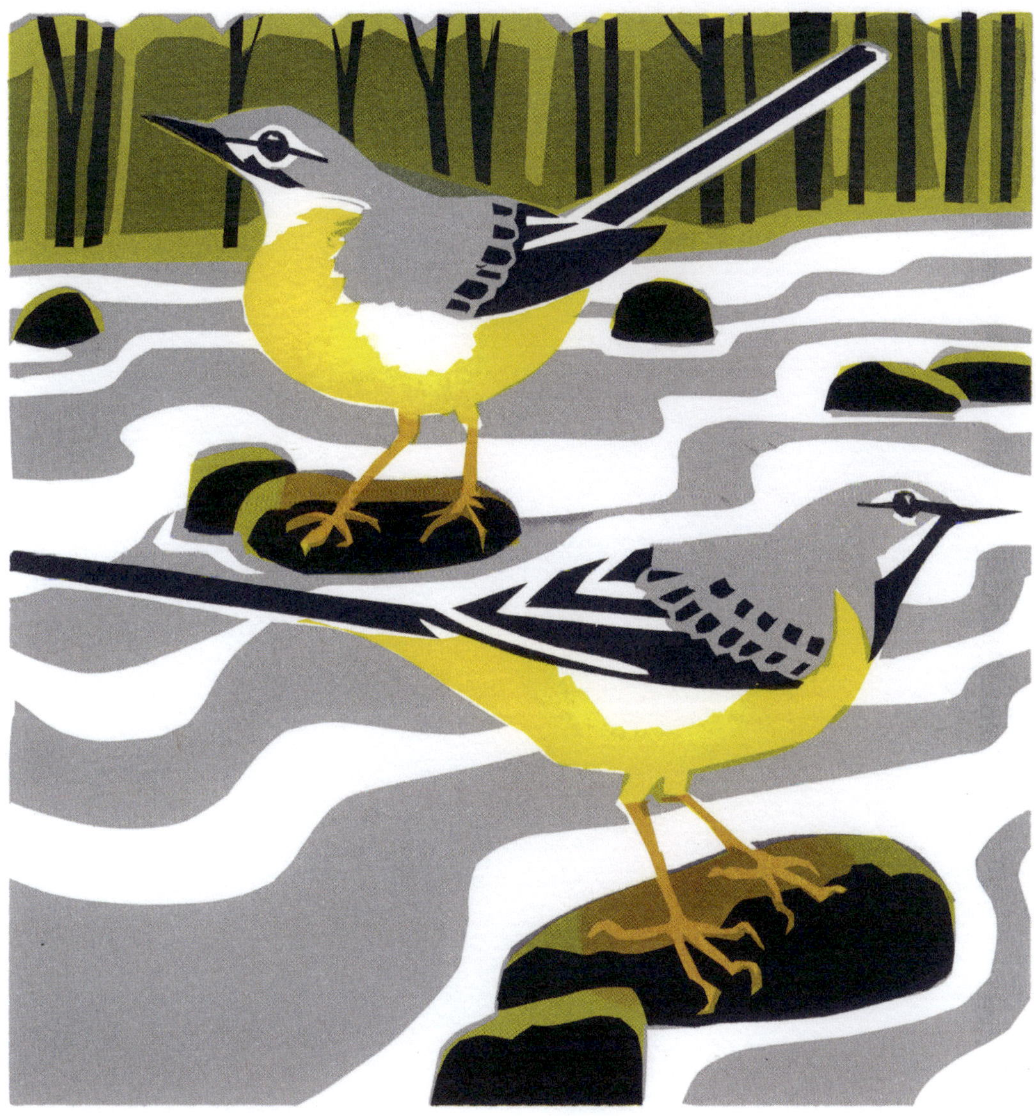

W. H. Hudson and Viscount Grey of Fallodon (1862–1933), politician and author of *The Charm of Birds*, were good friends; Grey even procured Hudson a pension. Paragons of bird writing, both described the **grey wagtail** (*Motacilla cinerea*), frequenter of rivers and streams, as exceptionally beautiful. The grey wagtail moults twice, in autumn and spring. In spring, the female's plumage brightens, and the male's white throat turns black. Gareth Thomas, photographer and Teme Weirs Manager, daily photographed a male that came to his bird feeder one winter, exceptional for an insect eater. The photographs suggested the male's colour changed not through moulting but within the feathers – a condition possibly also experienced by redshank, greenshank and king penguin.

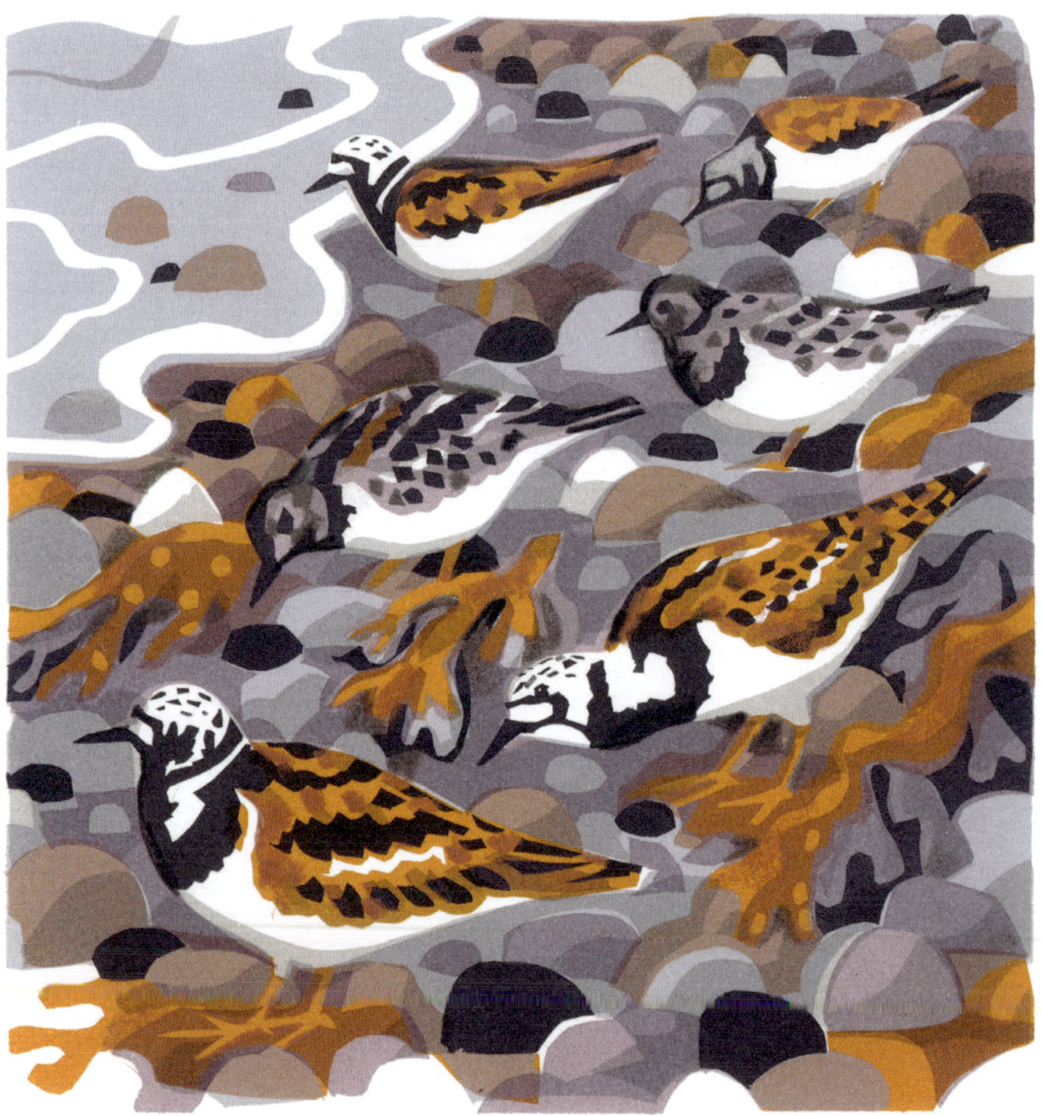

Waders are usually seen distantly, so it is a shock to find the **turnstone** (*Arenaria interpres*) running among seaside café tables or along a quayside in groups like scavenging starlings. By nature global wanderers, they migrate here at the end of the summer and are usually seen busily turning stones for food on the foreshore, sometimes a collaborative operation. As H. F. Witherby admitted in *The Handbook of British Birds*, 'a definite procedure in dealing with large objects seems indicated'. Duller plumaged in winter, late departures and early arrivals give glimpses of its tortoiseshell distinction.

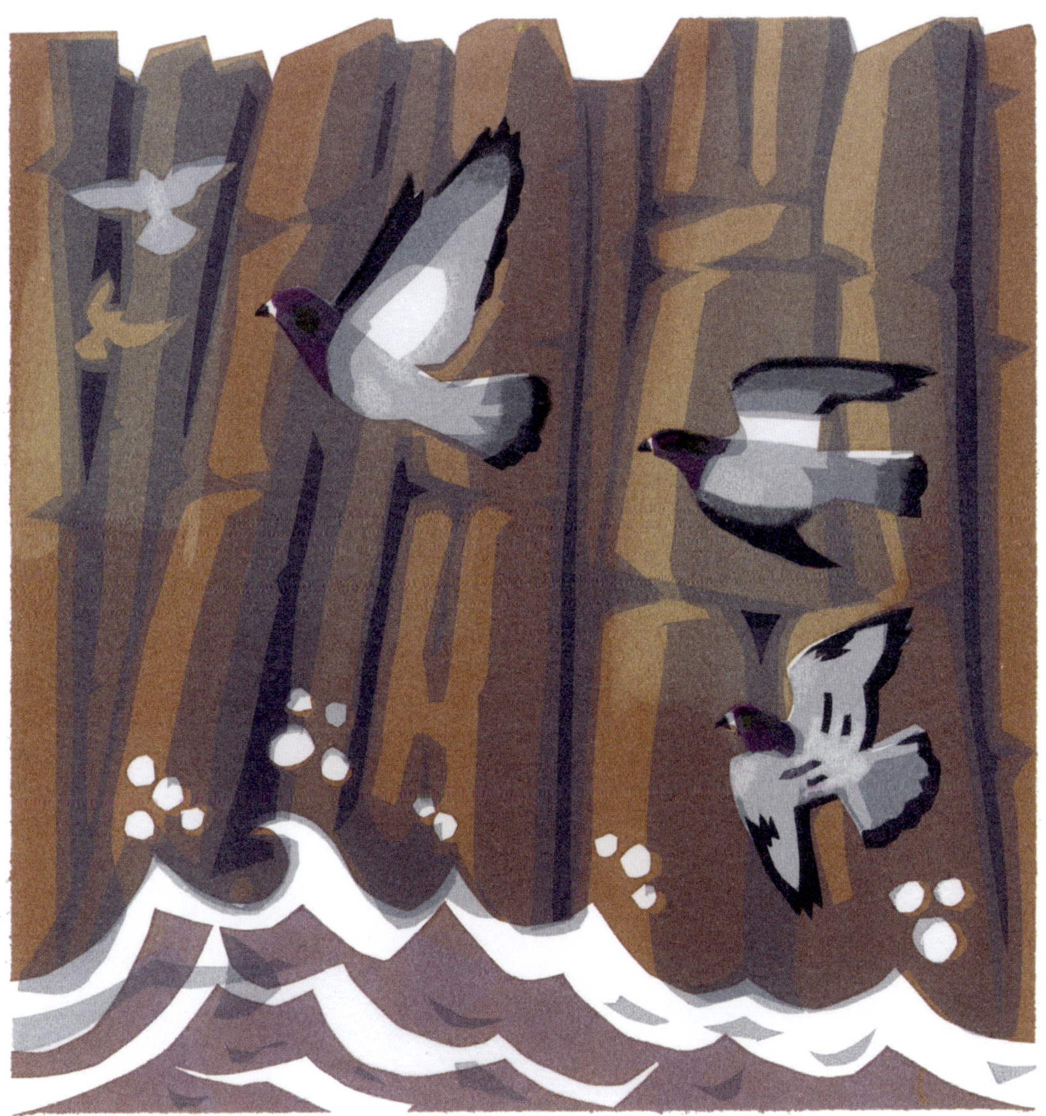

The **rock dove** (*Columba livia*), progenitor of pigeons, is threatened with extinction by hybridisation through mating with feral pigeons. This has already happened on the Isle of Man, though colonies on Shetland, Mull, Skye, Tiree and Islay so far show little sign of feral pigeon genes. Through selective breeding, the rock dove was moulded into the 228 distinct domestic forms named by Charles Darwin – most renowned of pigeon fanciers – from dovecote pigeons and racers to message-bearing carrier pigeons. Since 1943, carriers can win the Dickin Medal for valour and be buried with full military honours in a designated plot at the animal cemetery at Ilford, Essex. Escapee domestic pigeons gave rise to the city pigeon of feral in-distinction.

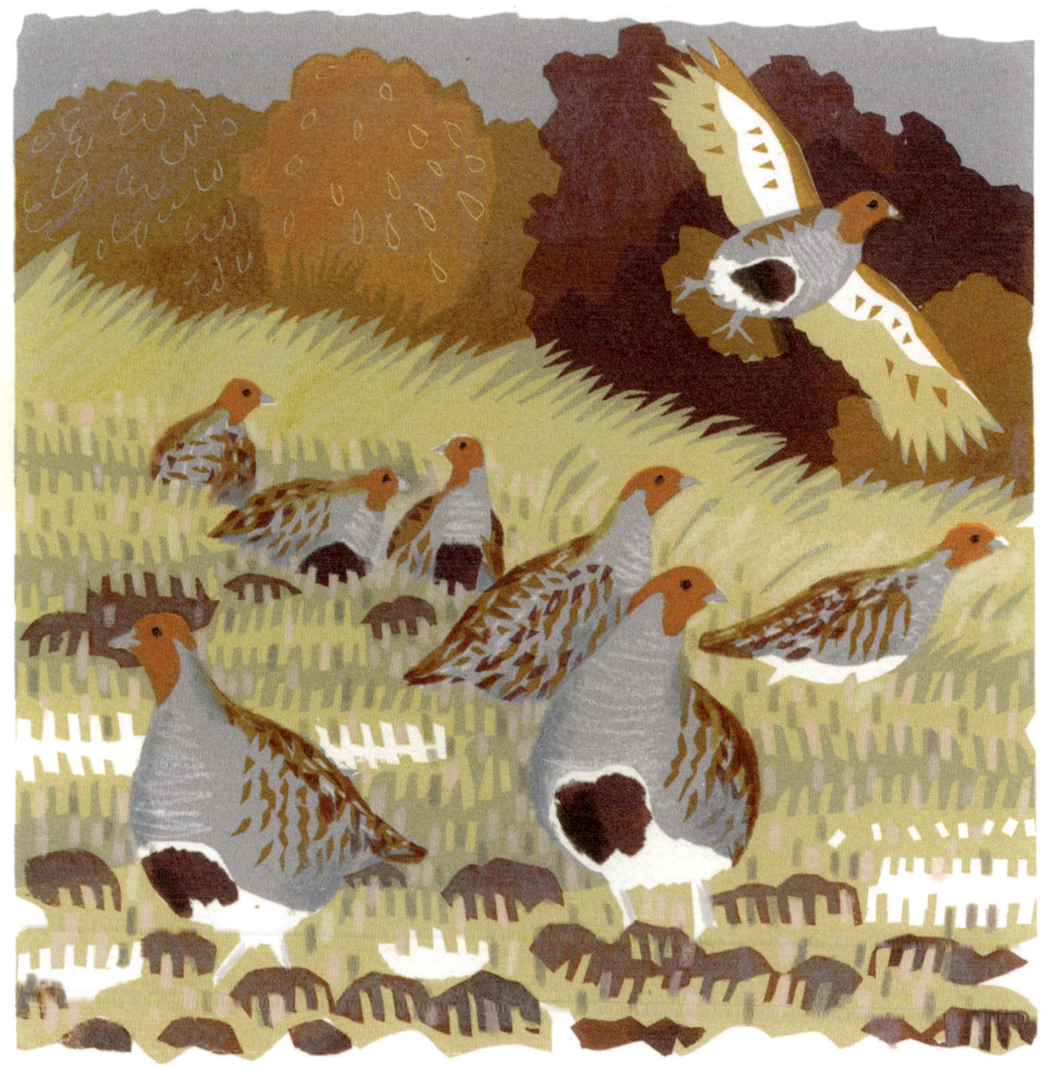

The **grey partridge** (*Perdix perdix*) is 'grey' only in recent UK bird guides, a demotion indicative of its headlong decline – 92 per cent since 1967. Without competition from other partridge species, it once reigned supreme. Loss of nest cover, seed and insect food through ever more intensive farming caused precipitous losses. We were warned. In September 1918, *Country Life* published a letter on the partridge from 'A.H': 'Nowadays farmers keep few objects more constantly before them than the extirpation of weeds and herbivorous insects. Their policy may be commendable from an agricultural point of view, but it means hunger and distress to the little brown birds.' Proliferating syndicate shoots replaced it with easier-reared red-legged partridges (10m released, 2019; grey partridge, 37,000 territories).

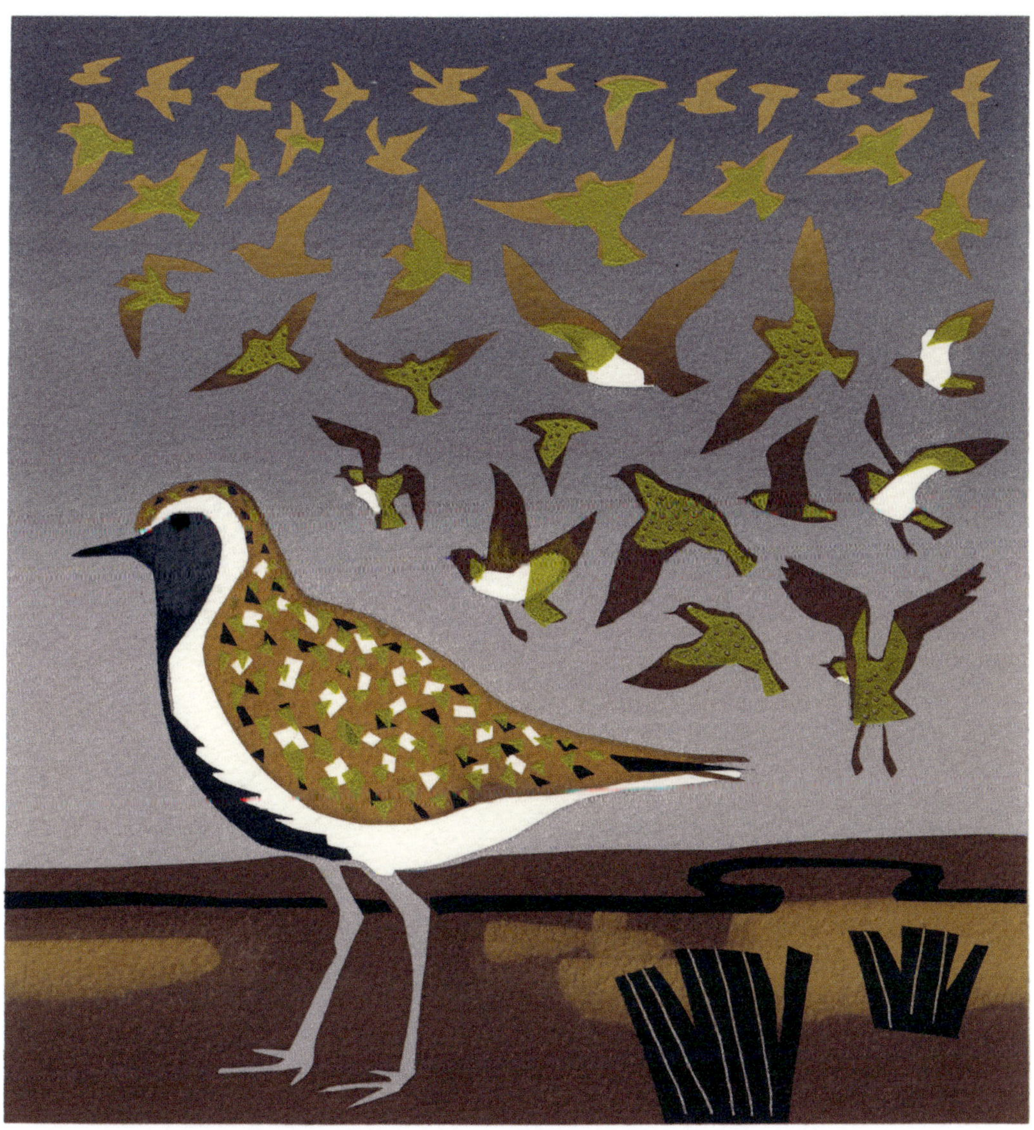

Golden autumn has its avian counterpart in the **golden plover** (*Pluvialis apricaria*, rainy/sunny, 'sunlit rain'). Like other waders, golden plover are at peak plumage in spring but at their most spectacular when they flock in autumn. 'Sunlit rain' perfectly describes waders in glimmering flight – the aerial equivalent of ocean shoals – especially the gilt flash of golden plover. In contrast, a nesting female within its moorland tussock is unflinchingly invisible from a couple of feet. Nineteenth-century naturalist and hunter Charles St John described them as 'a favourite prey' of the peregrine, but J. A. Baker (1926–87), author of *The Peregrine*, found 619 peregrine kills over 10 Essex winters, and only 2 per cent were golden plover.

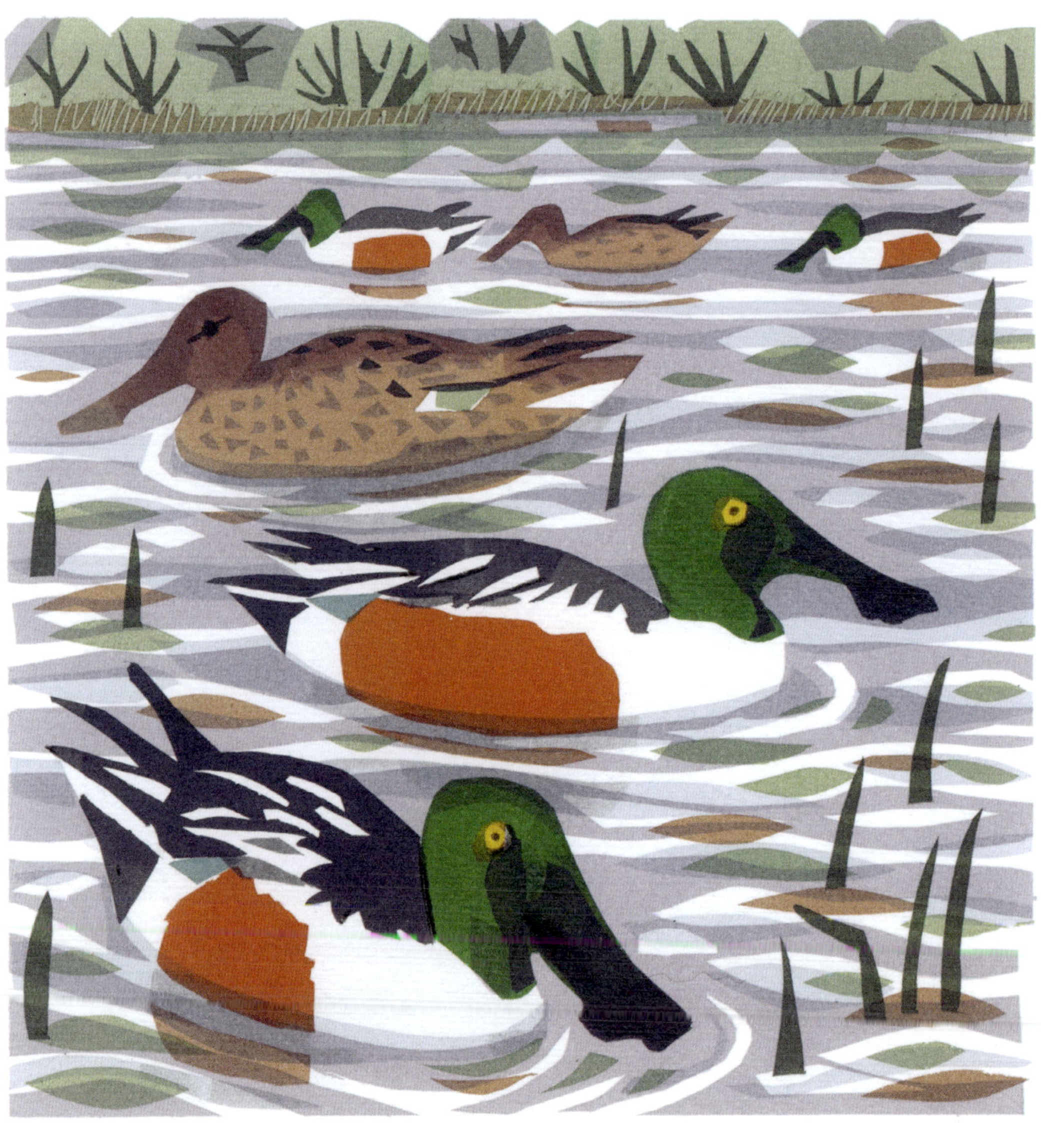

The spectacular colours of the drake shoveler (*Anas clypeata*) are the same as both sexes of shelduck. The spectacular male catches the eye and belies its modest numbers: 1,000 UK breeding pairs and 19,000 migrants in winter, which begin to arrive in September. At 10cm (4in), the bill is one-fifth of the bird's total length. The bill filters as the shoveler swims to feed, head extended and dabbling from side to side, often in comical circles to stir up food. There are 5m shovelers worldwide. Large quantities can be seen on the Nile, yet it does not appear in the tomb paintings of the ancient Egyptians.

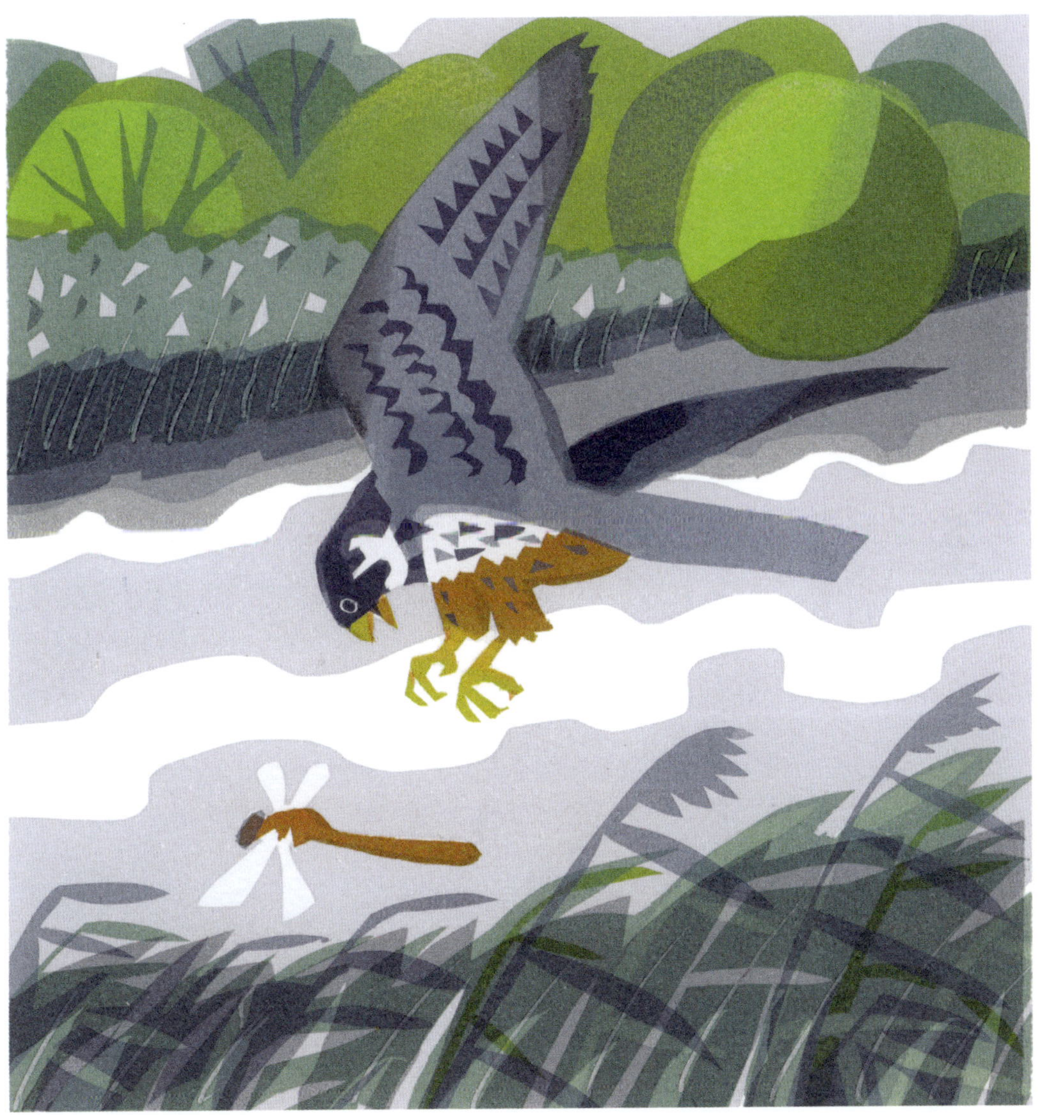

Peter Adolph (1916–1994), inventor of Subbuteo, the table-top football game, was not permitted to call his creation 'The **hobby**' (*Falco subbuteo*), so settled for its scientific epithet. A summer migrant from East Africa, the hobby's 100 pairs in 1970 were confined to southern England. Now, 2,050 pairs breed into the Highlands. Dragonflies are the hobby's favourite prey, but – described by ornithologist James Fisher (1912–1970) as the finest 'interceptor' – it can catch swifts. Migration finds flocks hunting over marshes. Field Marshal Alanbrooke (1883–1963), on wartime leave, 'retired to a wood in southern England and watched and filmed a pair of nesting hobbies' (*The Birdman*).

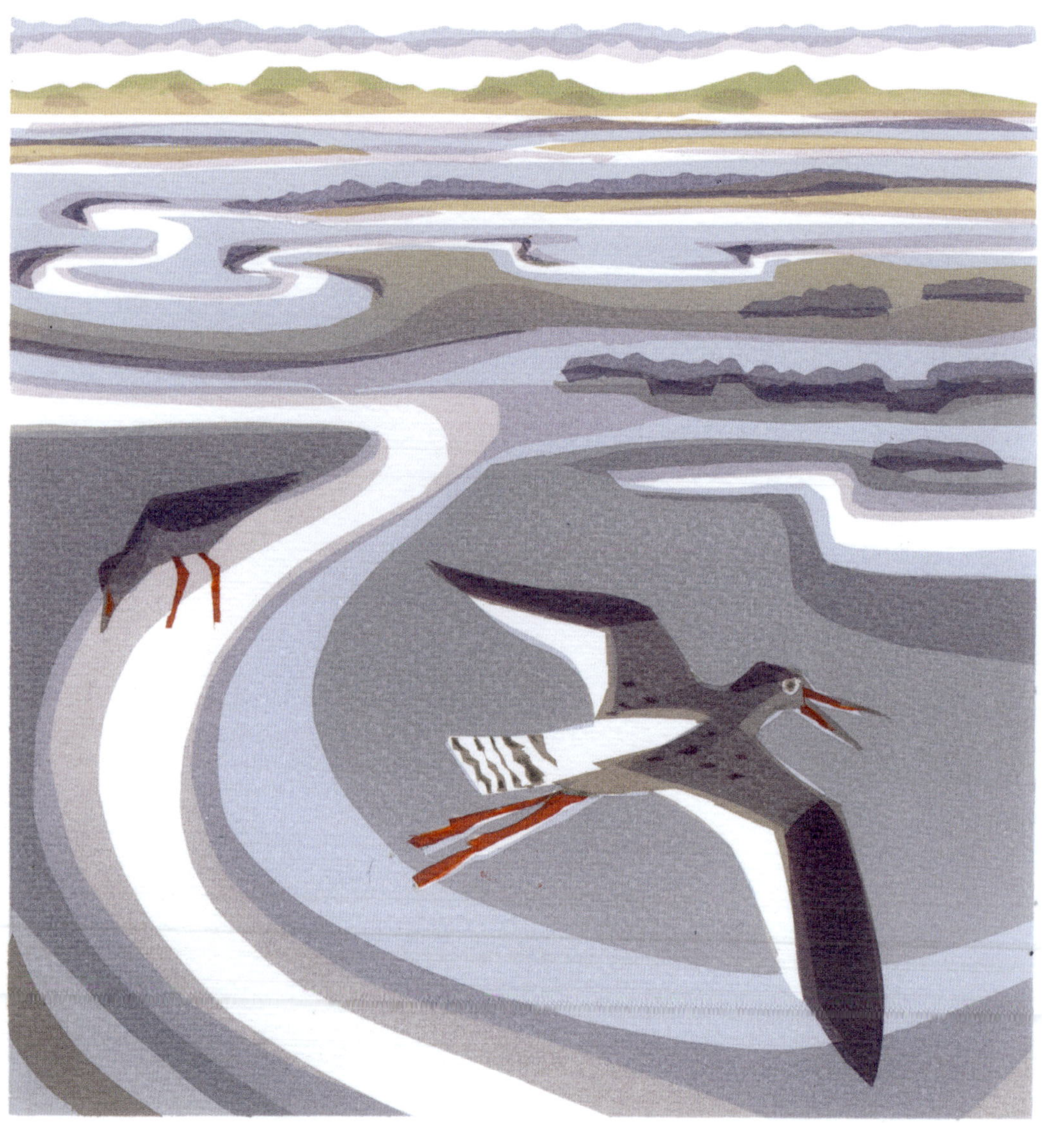

Among colloquial names for the **redshank** (*Tringa totanus*) are 'warden', 'yelper' and 'watchdog of the marshes', with good reason – its conspicuous departure and piping call invariably warn of visitors. The reduction of its breeding grounds is due, in the uplands, to deforestation. Not all felled trees are for timber: 7m trees are needed to meet the UK's demand for toilet rolls. The average Briton requires 127 toilet rolls (55km or 34 miles worth) each year. The figure doubles when disposable tissues and so on are included.

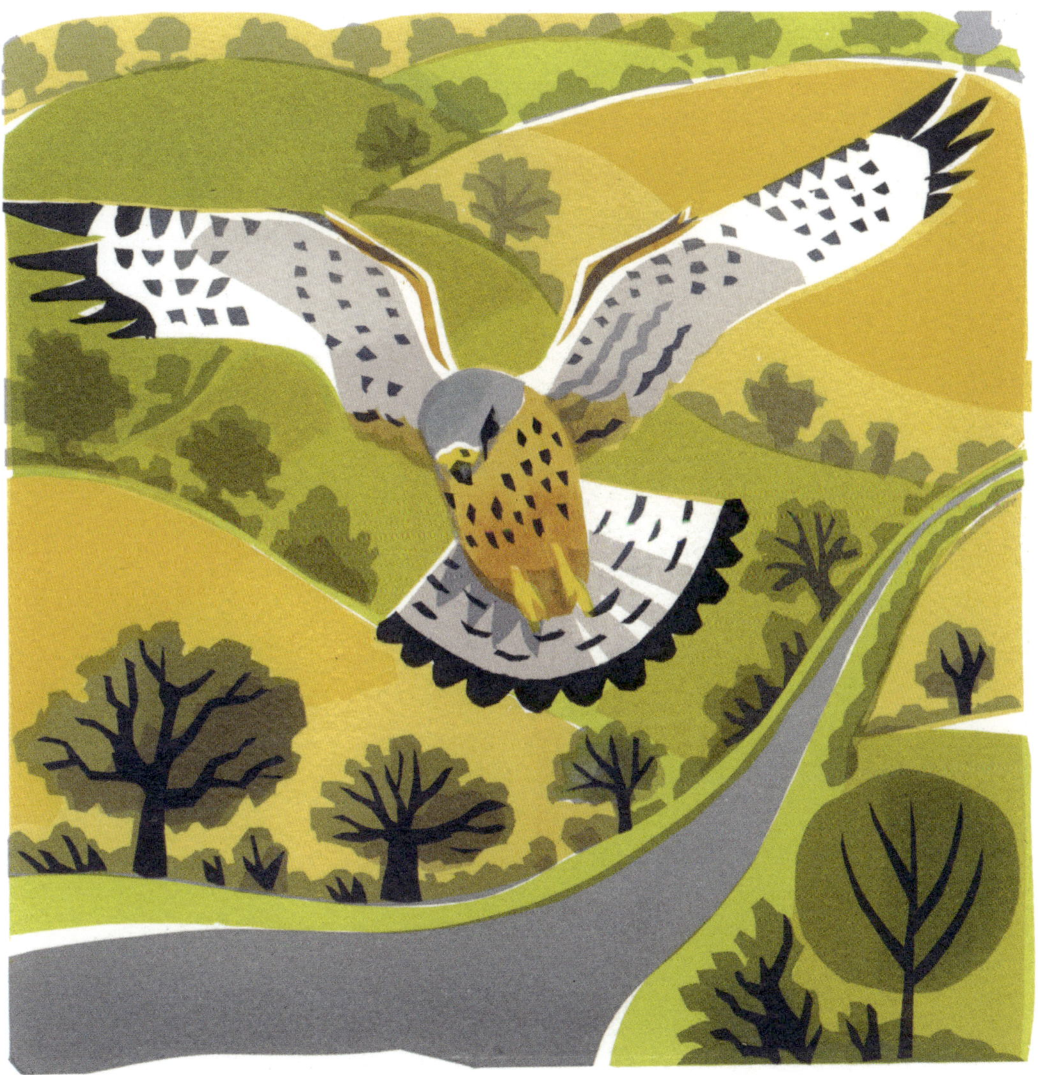

The **kestrel** (*Falco tinnunculus*) was the 'motorway falcon' until scrub growth on embankments aided its 42 per cent decline (from 1995–2021). Once our commonest raptor, it was the subject of Ken Loach's popular 1969 film *Kes*, from the 1968 novel *A Kestrel for a Knave* by Barry Hines (1939–2016), about a bullied boy finding joyful escape by adopting a kestrel. Its 30,000 pairs are now half those of the buzzard. The kestrel hovers to hunt, holding its position to within 1.25cm (0.5in), hence its old name, 'windhover'. Welsh Jesuit priest and poet Gerard Manley Hopkins (1844–1889) concluded 'The Windhover':

> *...My heart in hiding*
> *Stirred for a bird, – the achieve of, the mastery of the thing!*

Dedication: 'To Christ Our Lord'.

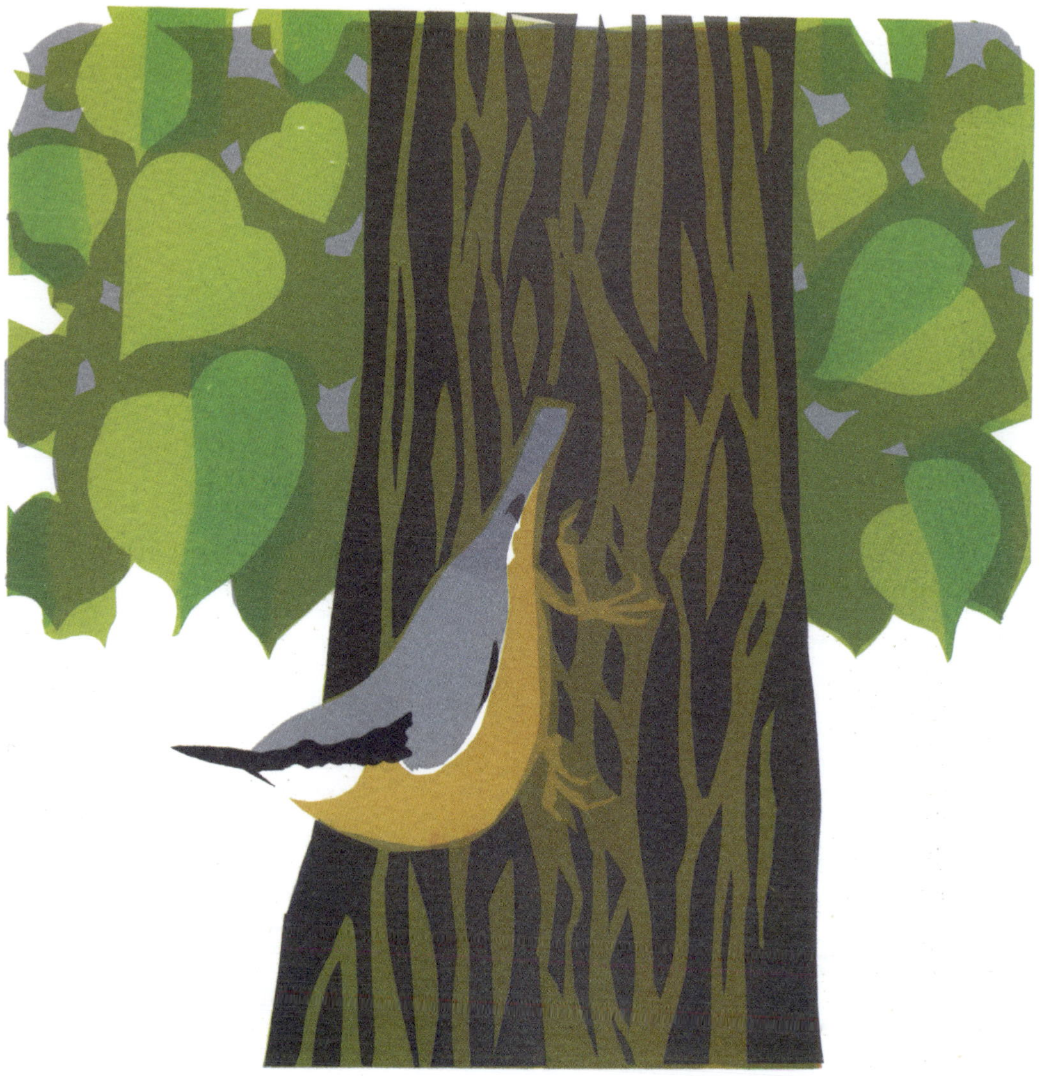

The **nuthatch** (*Sitta europaea*) is the only bird with a squirrel's ability to scurry down and around as well as up trees. Forthright in song and behaviour, in spring, the far-flung wolf-whistle of its call provokes cries of 'Who the heck's that!' Its formidable beak can noisily hack a wedged nut apart; its speedy flight similarly direct. Trees make it a regular visitor to garden bird feeders, and it can be hand fed – zooming in from a distance, taking a nut, and off in an instant. Breeding numbers have rocketed to 250,000 territories, and it has expanded its UK range, arriving in Scotland in 1989. Only Ireland is un-nuthatched.

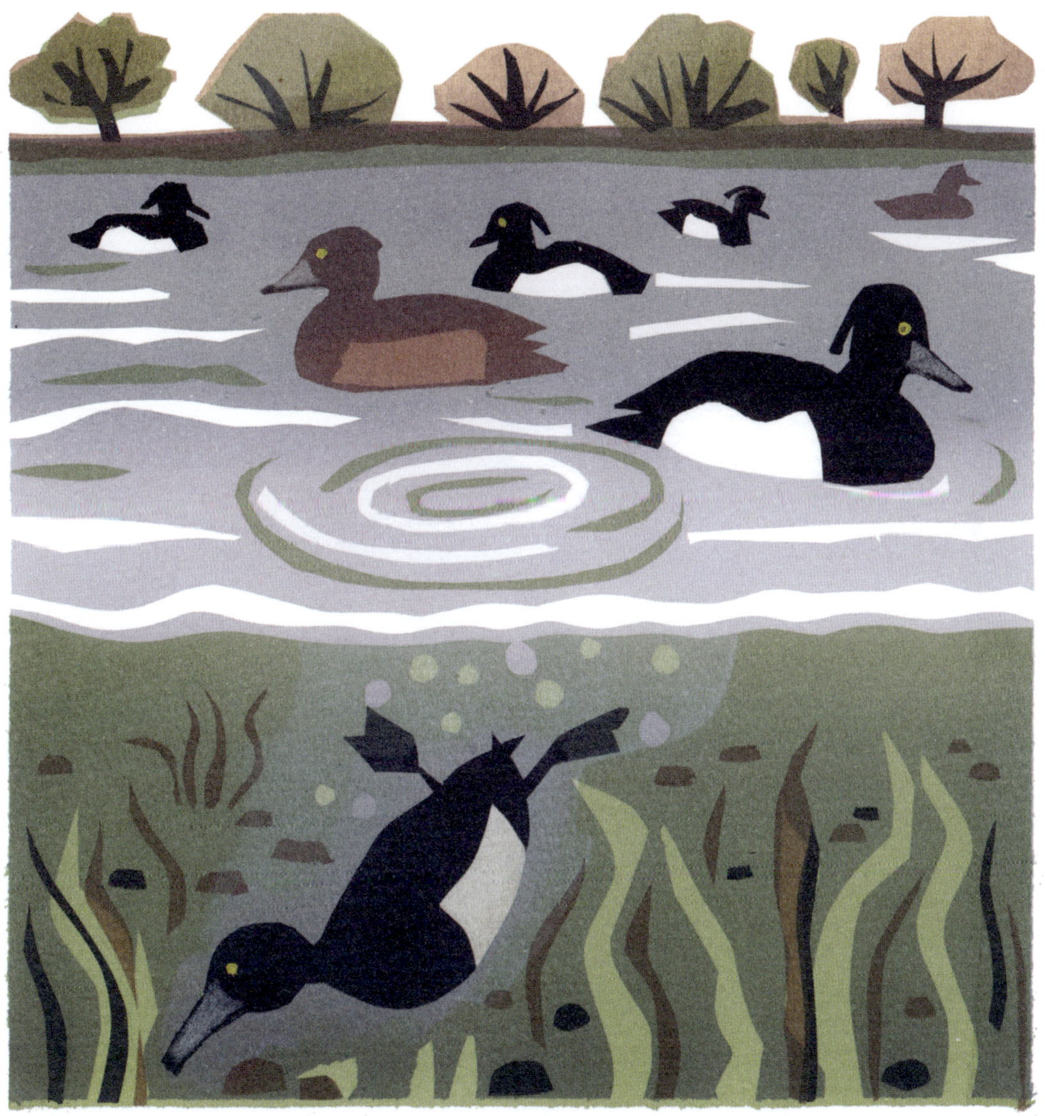

The drake's jaunty tuft, allied to the diving antics of the species, make the **tufted duck** (*Aythya fuligula*) a sure-fire children's entertainer. It was first recorded nesting in the UK in 1849 and today breeds widely, not least on urban lakes, the ducklings diving from birth. The winter influx of migrant tufteds (140,000 UK), mostly from Iceland, is the largest among freshwater diving ducks coming to the British Isles. Resident tufteds (19,000 pairs UK) can also migrate. A nestling ringed in Hertfordshire in May 1987 was found in Belarus three years later. Feeding underwater, they can use their bills to move quite large stones.

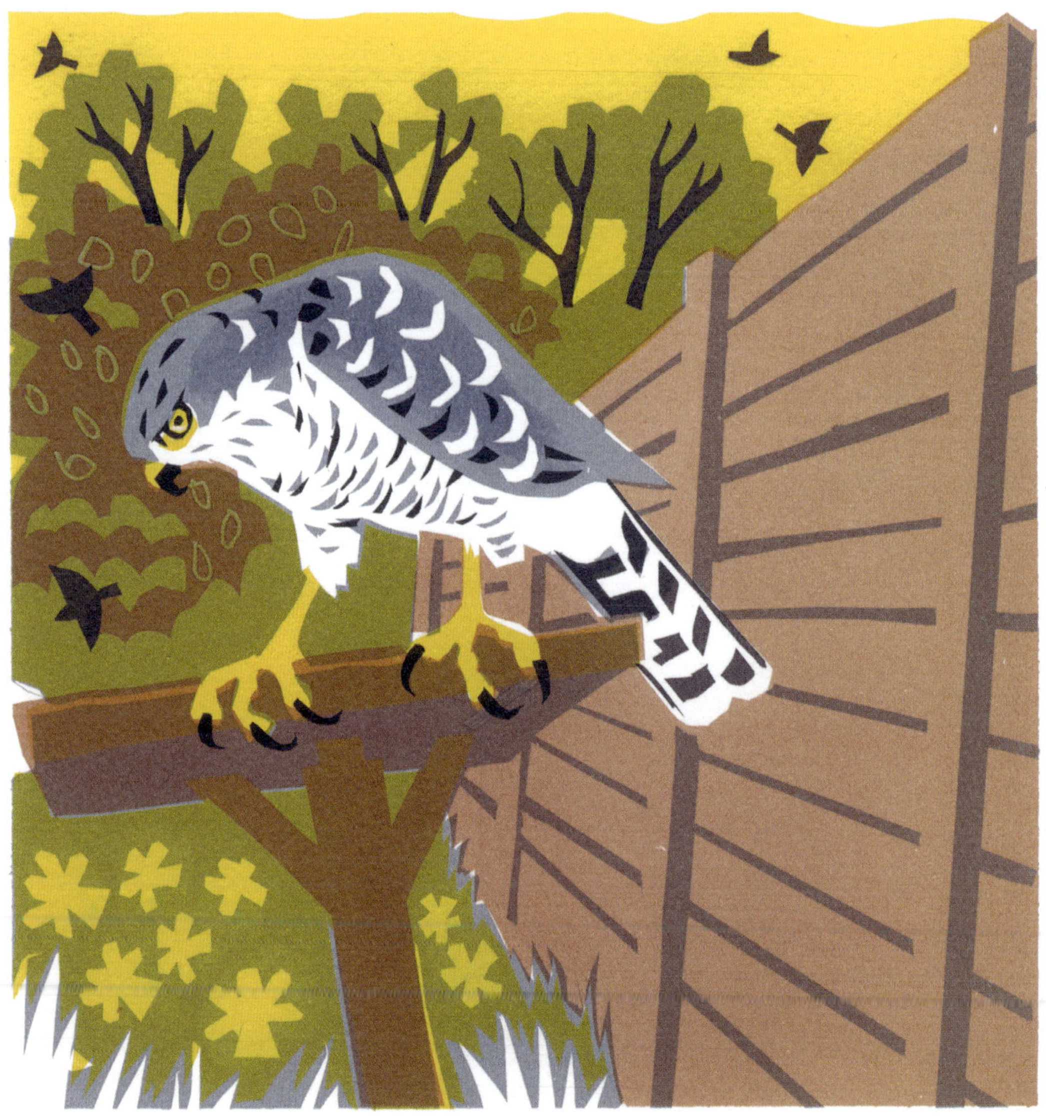

The **sparrowhawk** (*Accipiter nisus*), with the goshawk, is the smash-and-grab specialist among raptors. Its speed in ambush territory – woods, hedgerows, gardens – is electric. In the open it can still make a virtue of deception. One cruised overhead so nonchalantly I paid no attention to the settling woodpigeons in a field half a mile away. Looking again to check its progress, a barely discernible puff of white feathers was the answer. Bird tables are always imperilled. The sparrowhawk has long been protected, helping it to recover somewhat from the damaging impact of pesticides. Its estimated 30,500 UK pairs raises calls for control from some, though its population has fallen by 25 per cent in recent years (1995–2020).

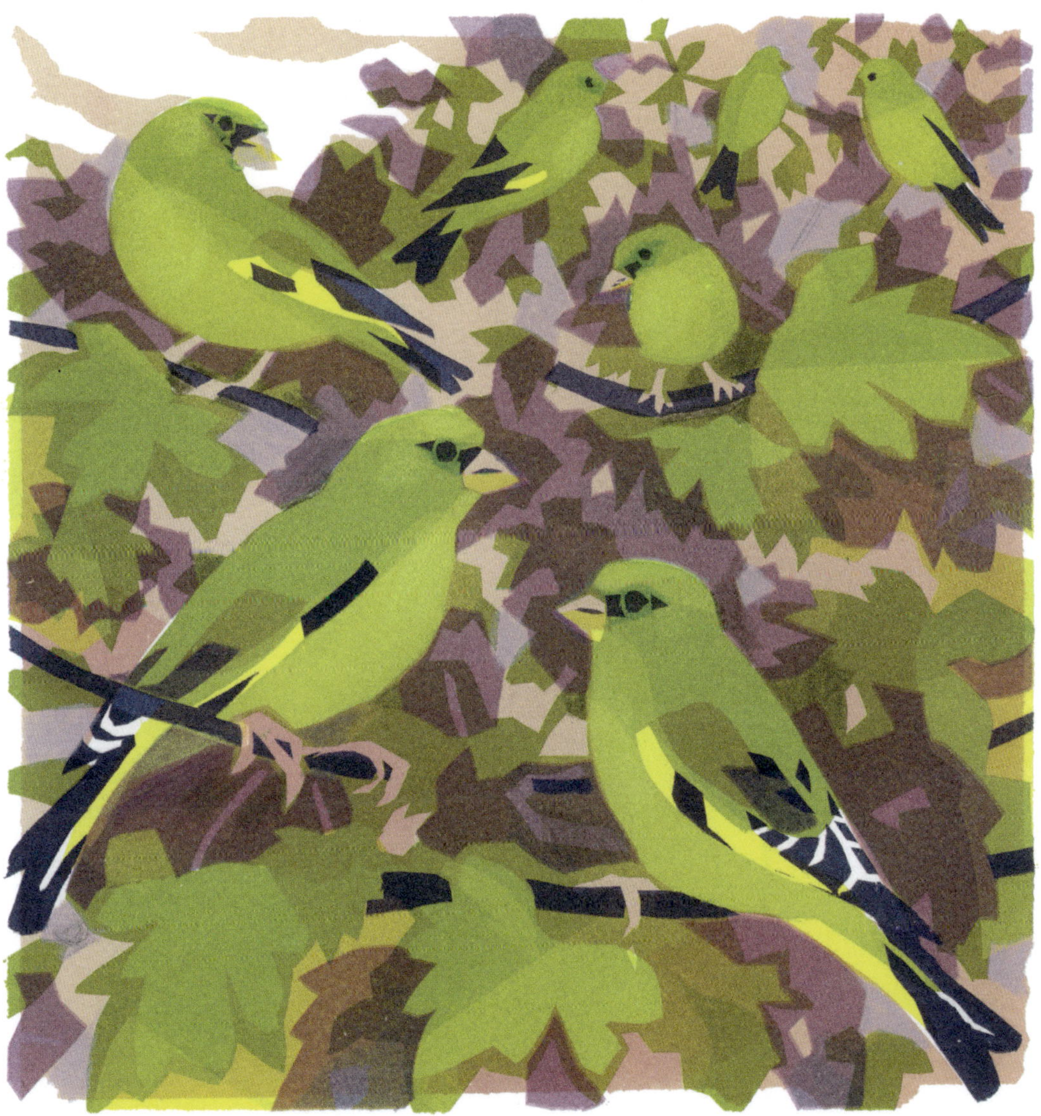

The **greenfinch** (*Chloris chloris*) has a ditty but is most easily recognised by its repeated call of *cheese*. Over recent decades it has fallen victim to a protozoal parasite. Parents feeding contaminated food to their young was one cause, with dirty bird feeders and bird baths also to blame. The parasite is not transmitted to humans. Although the greenfinch's 69 per cent decline since 1967 is drastic, there are still 785,000 UK pairs.

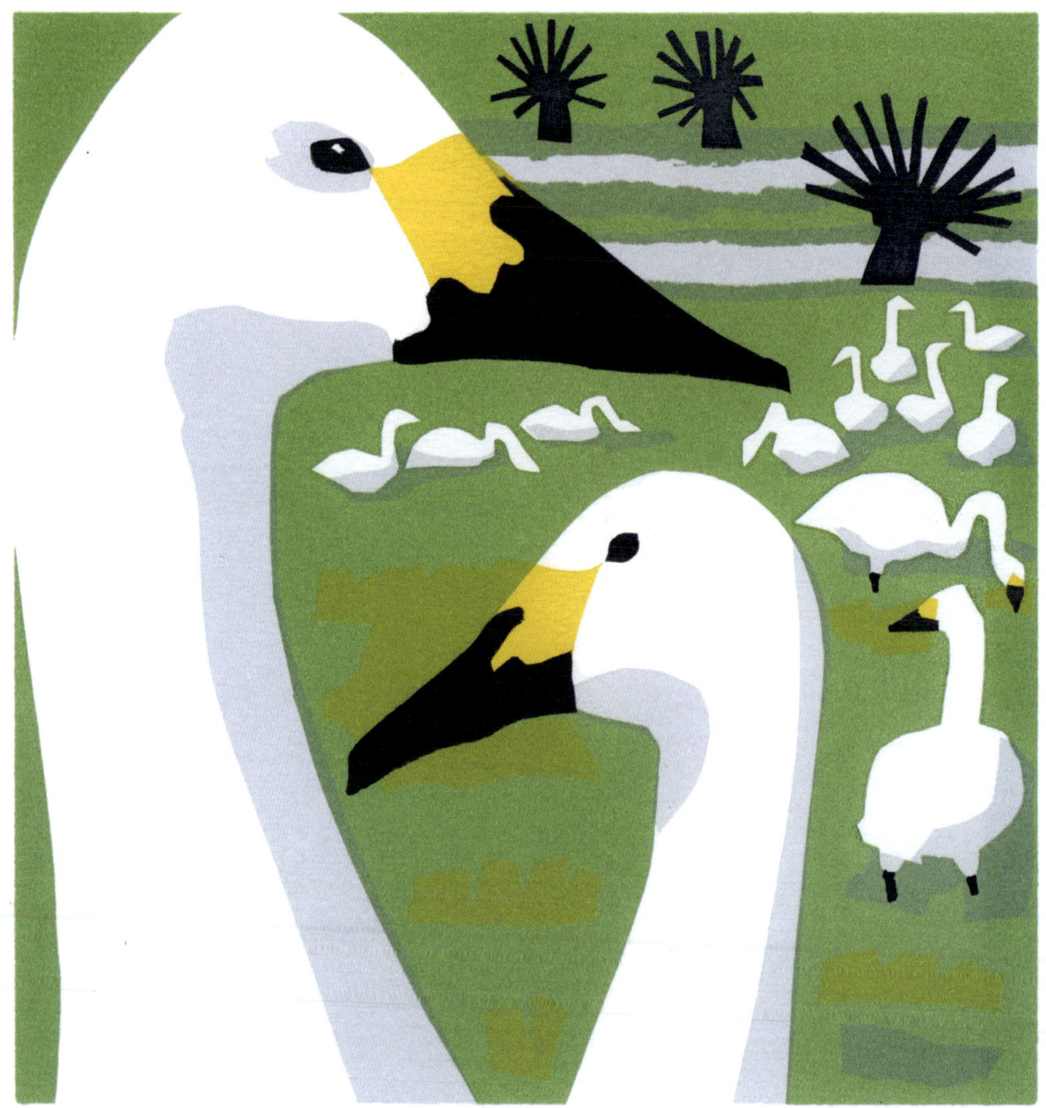

In Northumberland in 1830, the ornithologist William Yarrell (1784–1856) identified a new swan species and named it **Bewick's swan** (*Cygnus bewickii*), after Northumbrian artist and pioneering author and engraver of *A History of British Birds* Thomas Bewick (1753–1828). In 1964, artist and conservationist Peter Scott (1909–1987), founder of the Wildfowl & Wetlands Trust (WWT) at Slimbridge, Gloucestershire, noticed that each Bewick's swan's black-and-yellow bill pattern is unique; it remains the only avian record not ring- or tracker-dependent. Their UK migration has declined to 1,278 (2020) from 7,100 (2010). In 2016, Sacha Dench (working with the WWT) accompanied migrating Bewick's swans by paramotor the 7,001km (4,350 miles) from Siberia to England. A Slimbridge treasure is Bewick's telescope.

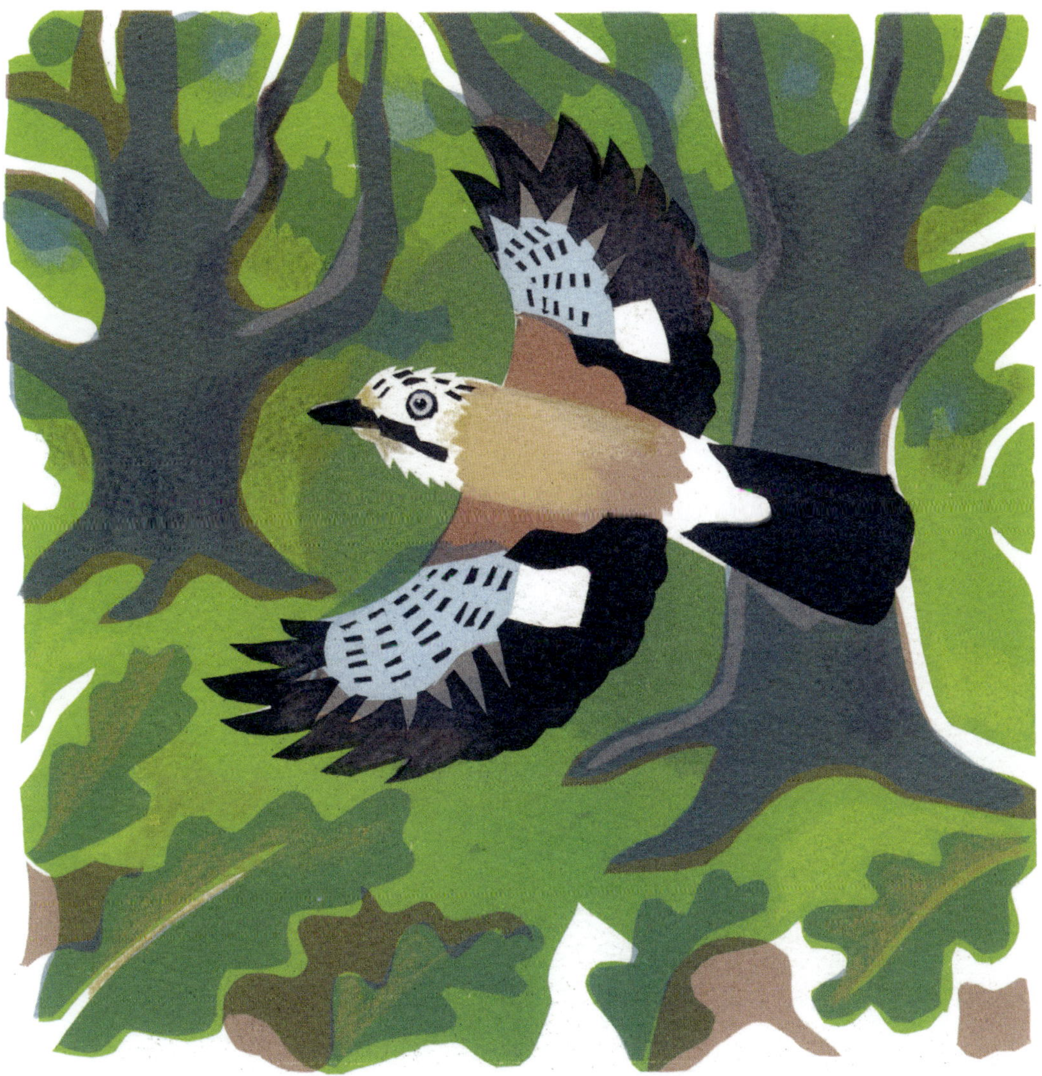

The **jay** (*Garrulus glandarius*) is an omnivore; its summer diet includes the eggs and chicks of smaller birds. In autumn, it feasts on acorns, burying the surplus nuts for after the harvest. Those that are not recovered might grow into oak trees. Oaks supplied the wood to build the ships of the Royal Navy. It took 5,000 oak trees (the annual number of acorns buried by a jay) to build Trafalgar veteran the 'fighting' *Temeraire*, immortalised by J. M. W. Turner's (1775–1851) painting. In 1880, the scandal caused by the Duchess of Edinburgh flaunting a sky-blue muff of jay feathers provoked Emily Williamson to start the movement for bird protection that became the RSPB.

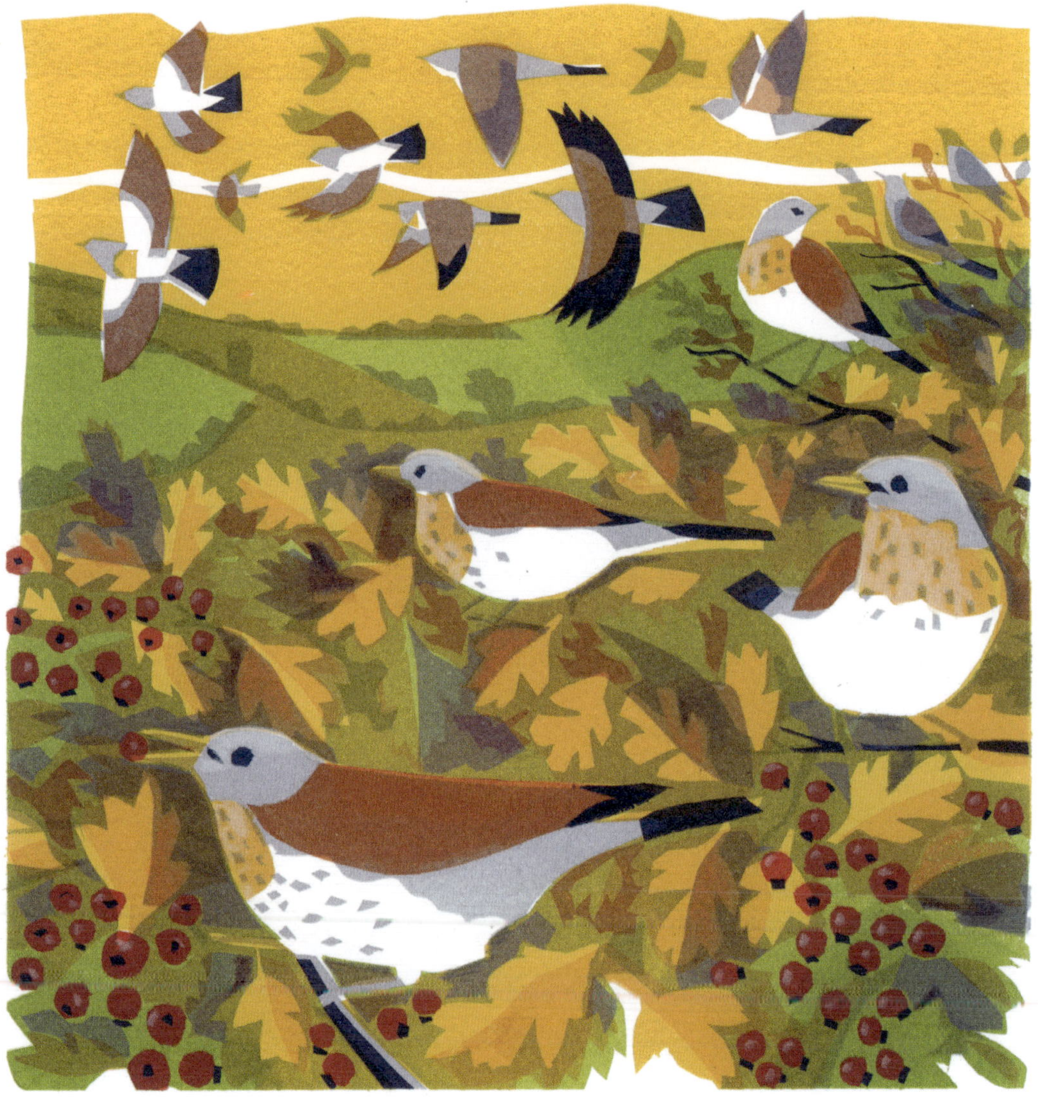

In October, the UK's 730,000-strong winter migration of the **fieldfare** (*Turdus pilaris*) from Fennoscandia (Norway to Russia) is in full swing. Many bird species 'fare' (travel) over fields, the fieldfare grazing for worms or plundering the berry harvest in the hedges. Only a prolonged freeze brings them to town where their size makes them dominant contestants for food. A very rare breeding bird in the UK, fewer than five pairs have been reported in most years since the turn of the century.

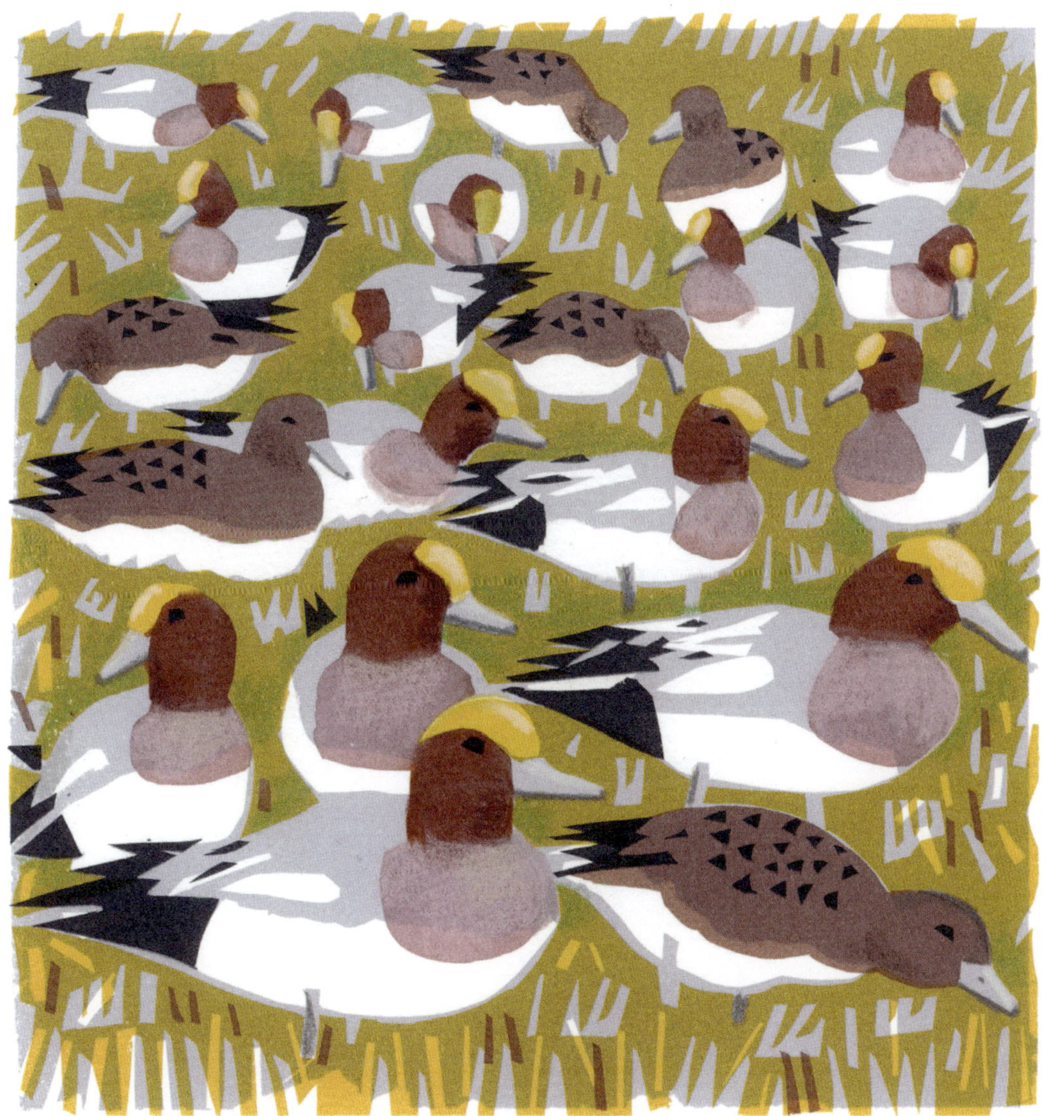

There are 200 resident **wigeon** (*Mareca penelope*) pairs in the UK, whereas 450,000 individuals migrate here in winter. They are grazing ducks, seen by open inland waters and on marshes everywhere. Although associated with the sea and foreshore, they can turn up for a passing visit to the merest water flash in a field and will spend whole winters on inland reservoirs and lakes. At Elmley Nature Reserve, Isle of Sheppey, rentable accommodation affords overnight proximity to plentiful winter packs and their lonesome whistling.

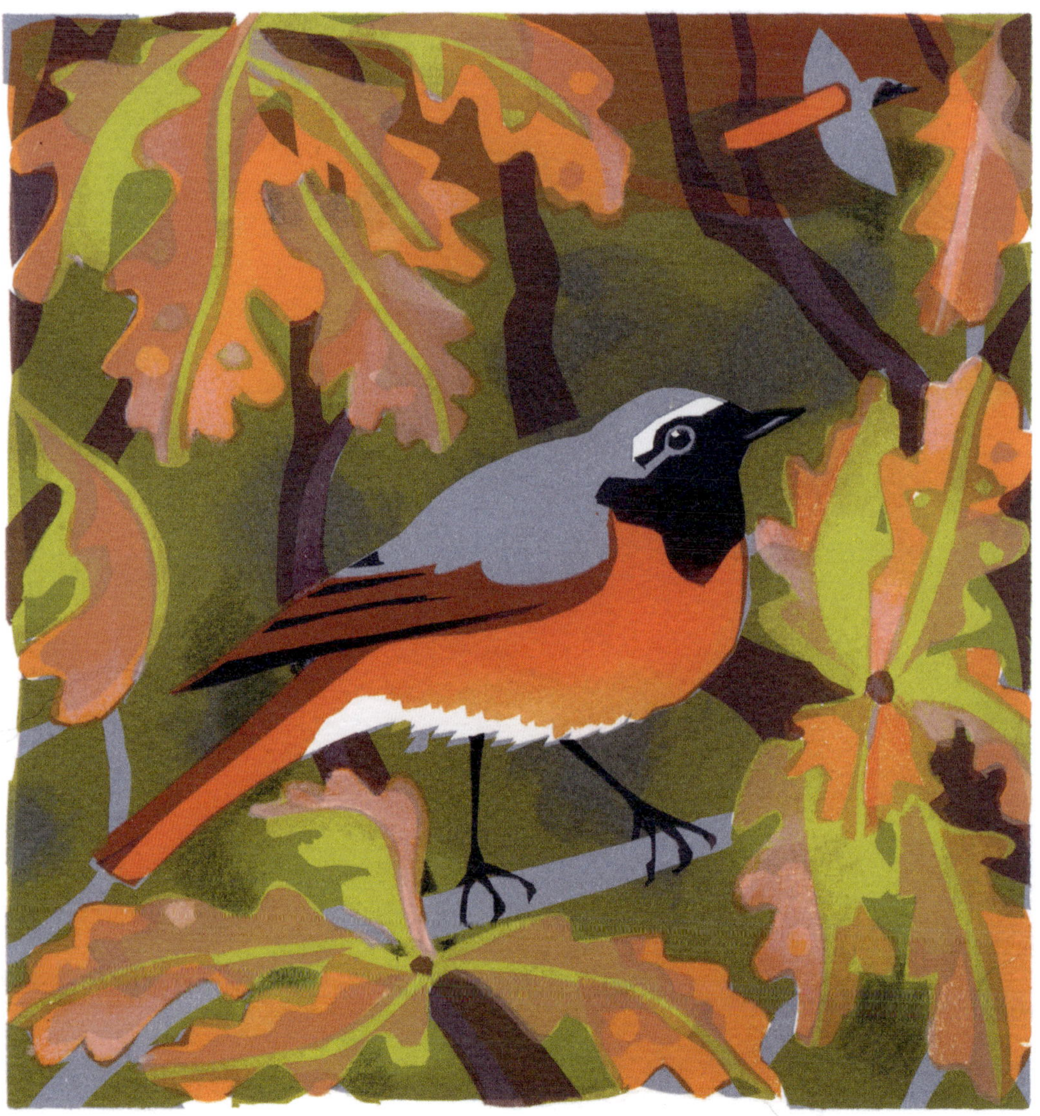

When John Buxton (1912–1989) was a POW in rural Germany in 1943, his brother-in-law, Ronald Lockley, advised him to birdwatch. Buxton chose a migrant **redstart** (*Phoenicurus phoenicurus*) pair. Buxton and fellow camp bird lovers were liberated, the redstarts 'unconcerned by our fatuous politics'. The result was Buxton's post-war bird classic, *The Redstart.* The name *Phoenicurus*, 'firetail', was bestowed by Aristotle. Four hundred of its rare cousin, the black redstart (*Phoenicurus ochruros*), winter here. It likes rubble and first arrived in an appreciable number to make the most of war-blitzed London. There are up to 44 breeding pairs, the majority in London.

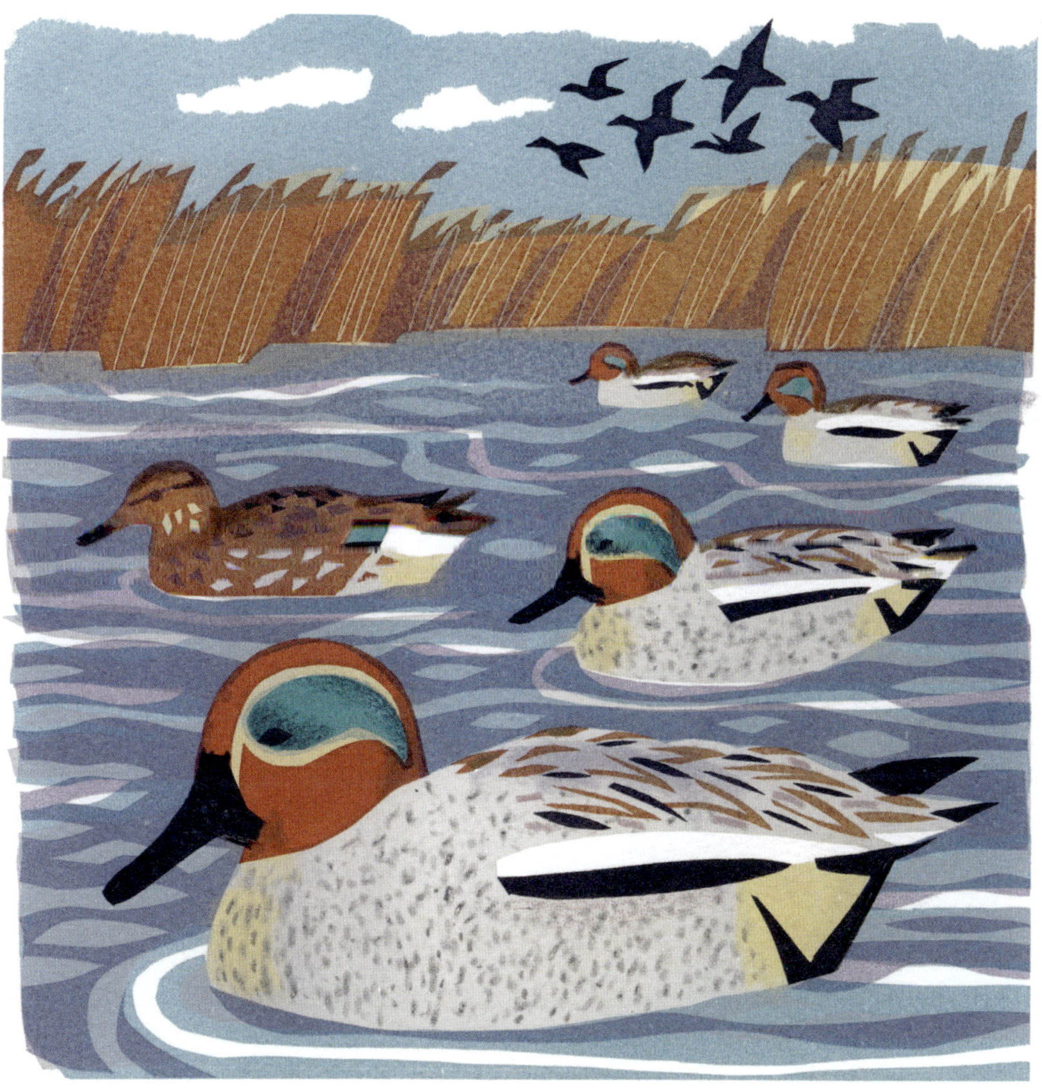

The **teal** (*Anas crecca*) is on the increase: 4,500 resident UK pairs are joined in winter by 1.5m migrants from Europe and beyond. The smallest of British ducks, the drake is also among the most beautiful. The green bar of colour on the wing of both sexes, the speculum (Latin for 'mirror'), is well-named – sunlit, it can blaze like a traffic light on 'go'. We say a 'spring' of teal, which this fast flyer does when disturbed, the female often first to jump.

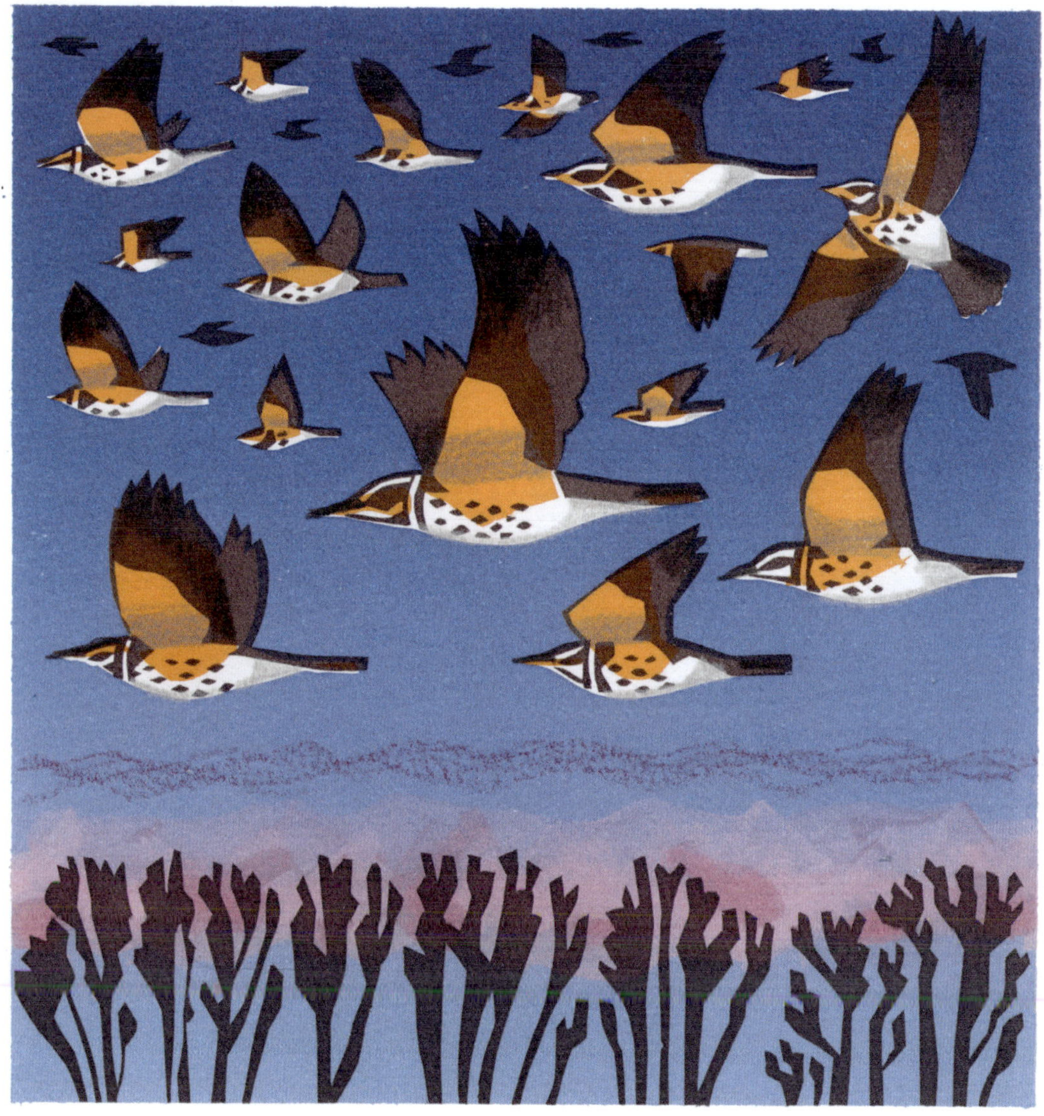

The **redwing** (*Turdus iliacus*) is as gentle as its soft *see-ap* call, in contrast to the *chuk-chuk* of its aggressive, larger thrush cousin, the fieldfare. Redwings are the thrushes most vulnerable to harsh winters. In the winter of 1962–1963, almost the entire British migration died. In his memoir, *A Diplomat Off Duty*, Sir Francis Lindley (1872–1950) wrote that the redwing symbolised the Norwegian spring – its loud, three- or four-note song, heard wherever there were trees: 'All day, and nearly all the light night'. He wondered why it did not breed in Britain. Annually, about 20 pairs take his advice, all in north Scotland.

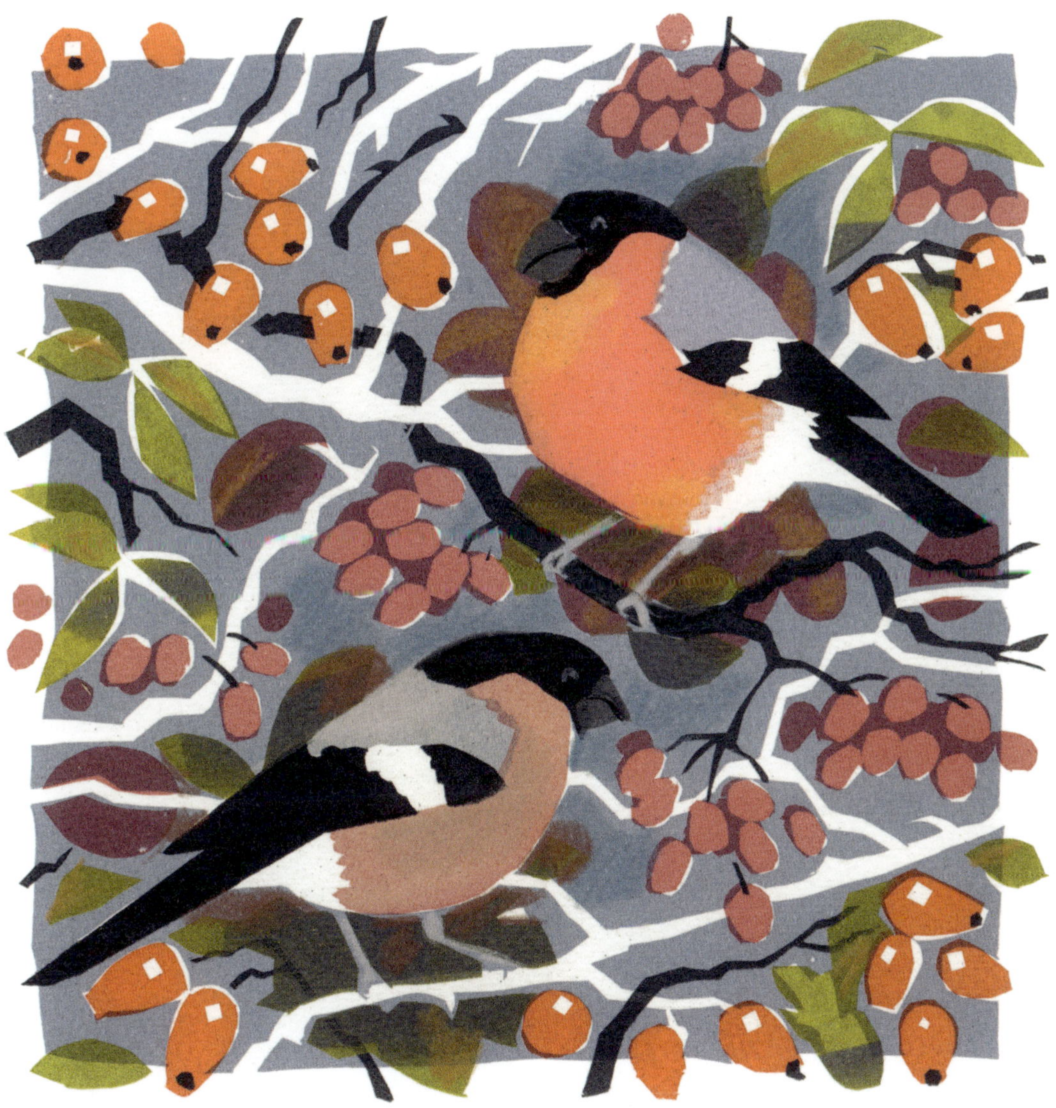

Denys Watkins-Pitchford (BB) thought there was no prettier sight than a **bullfinch** (*Pyrrhula pyrrhula*) pair – bullfinches tend to be in pairs – perched among rosehips in an autumn hedgerow of his native Northamptonshire. It was a subject he never tired of painting. The bullfinch is a country bird reduced in numbers, yet its taste for fruit tree buds once made it such an enemy of cherry farmers that into the 1980s, it was still legally shot in England to protect some orchards. BB saved abandoned bullfinch nestlings. If he settled in an armchair, his favourite, Biddy, would snuggle in his hair for a nap.

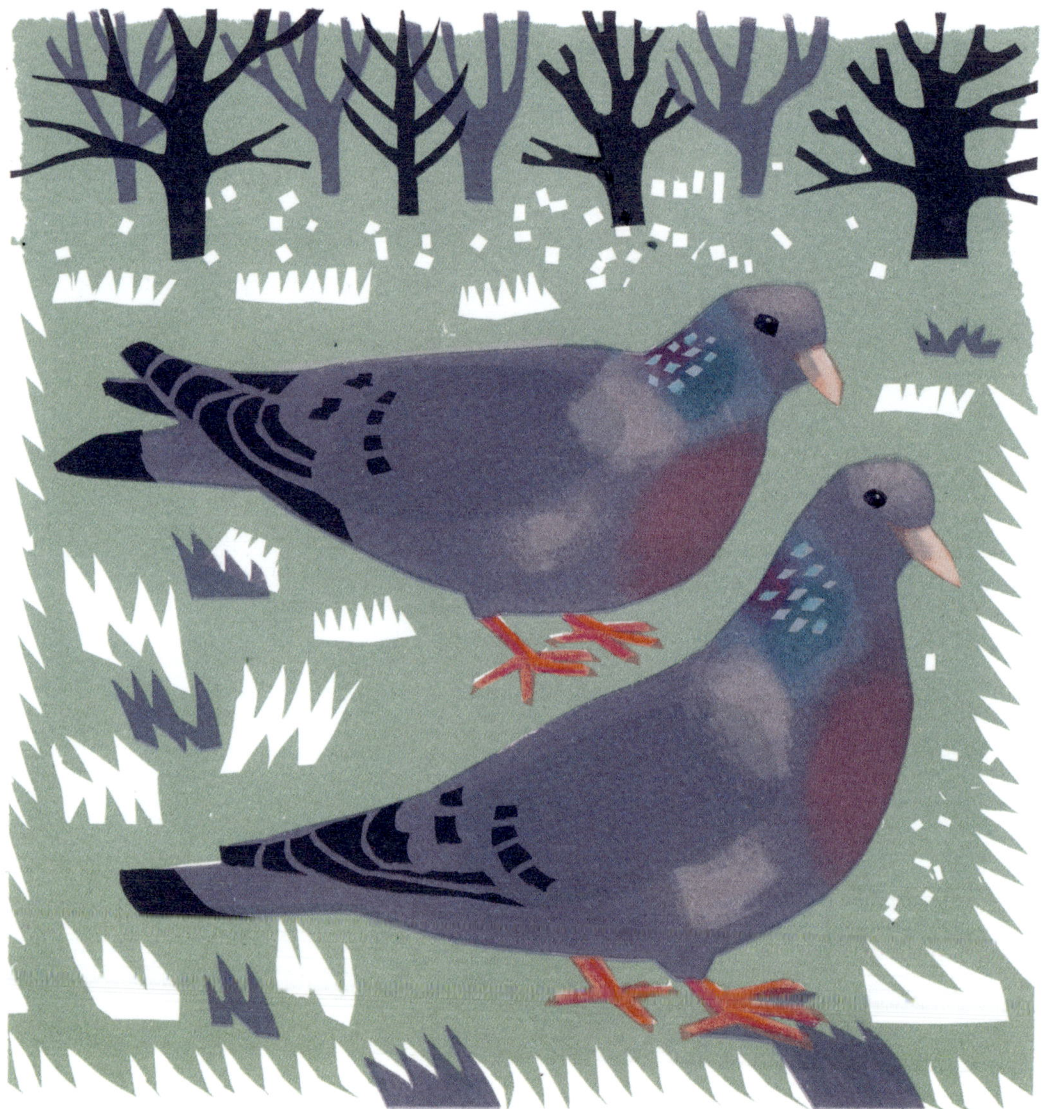

The **stock dove** (*Columba oenas*) was not scientifically declared a separate species to the woodpigeon until the early 1880s. The earliest known reference in English is 'stok-dowe' (1340). 'Stock', which means 'tree trunk', refers to its nesting in tree holes, unlike branch-nesting woodpigeons. Its two-note call, *oo-waa, oo-waa*, is repetitive and upbeat:

> *The birds are singing in the woods*
> *Over his own sweet voice the Stock-dove broods*
>
> William Wordsworth, from 'Resolution and Independence'

In the absence of tree holes, stock doves will nest in burrows. They retain the wildness of a country bird, even if nesting in city parks, not associating with feral pigeons.

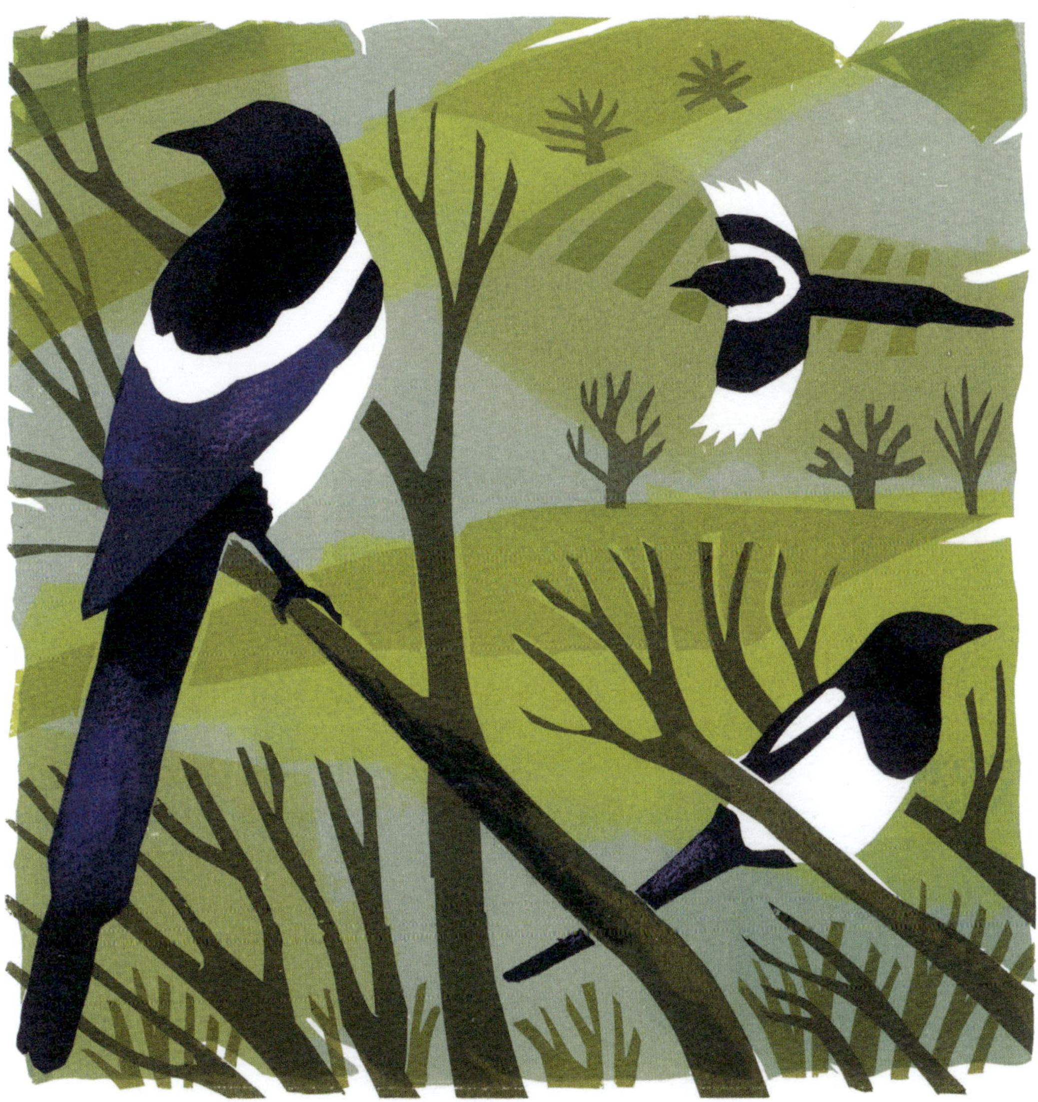

The **magpie** (*Pica pica*), being black and white (*pie*bald), is said to be cursed with a blood drop of Satan, Prince of Darkness. Hats are doffed at the sight of one to ward off bad luck. A magpie group is a 'mischief'. One magpie rhyme goes:

One for sorrow, two for mirth,
Three for a funeral, four for a birth

'Sorrow' because they are monogamous. *The Thieving Magpie* opera by Rossini (1792–1868) immortalises the birds' fabled stealing – much exaggerated though once a platinum engagement ring was found in a nest. I once prevented a young starling from being killed by first a magpie, and then a cat, which rushed up hoping to finish the job – UK cats kill 27m birds a year, mostly fledglings.

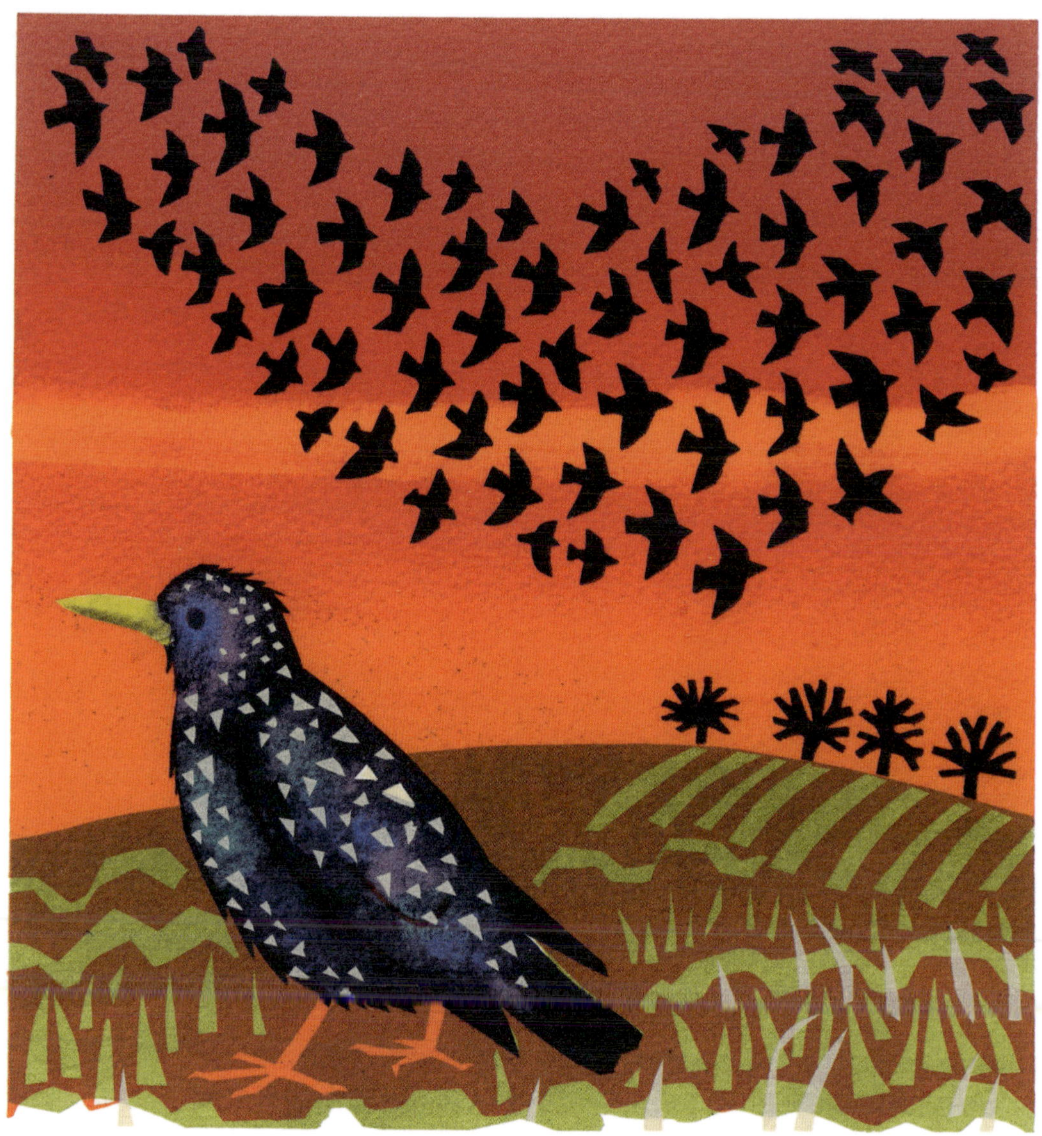

The **starling** (*Sturnus vulgaris*) is renowned as a mimic – even of the smoker's cough of bird artist James McEwen (1924–71) down his bedroom chimney. In *A Sentimental Journey* by Rev. Laurence Sterne (1713–68), the caged starling, embodying 'the miseries of confinement', repeatedly complains, 'I can't get out'. On 27 May 1784, Mozart (1756–91) – first symphony written aged eight, in London – bought a caged starling, perhaps because it sang some notes of his music. When it died, he buried it with full funereal honours. His *Divertimento for two horns and string quartet, K.522*, classified 'A Musical Joke', is thought a parody of its random mimicry. Giorgio Parisi (1948–), author of *In a Flight of Starlings*, was fascinated from childhood by Rome's swirling winter starling flocks. His youthful discovery of a pattern to their apparent randomness led to a career that earned him the 2021 Nobel Prize for Physics.

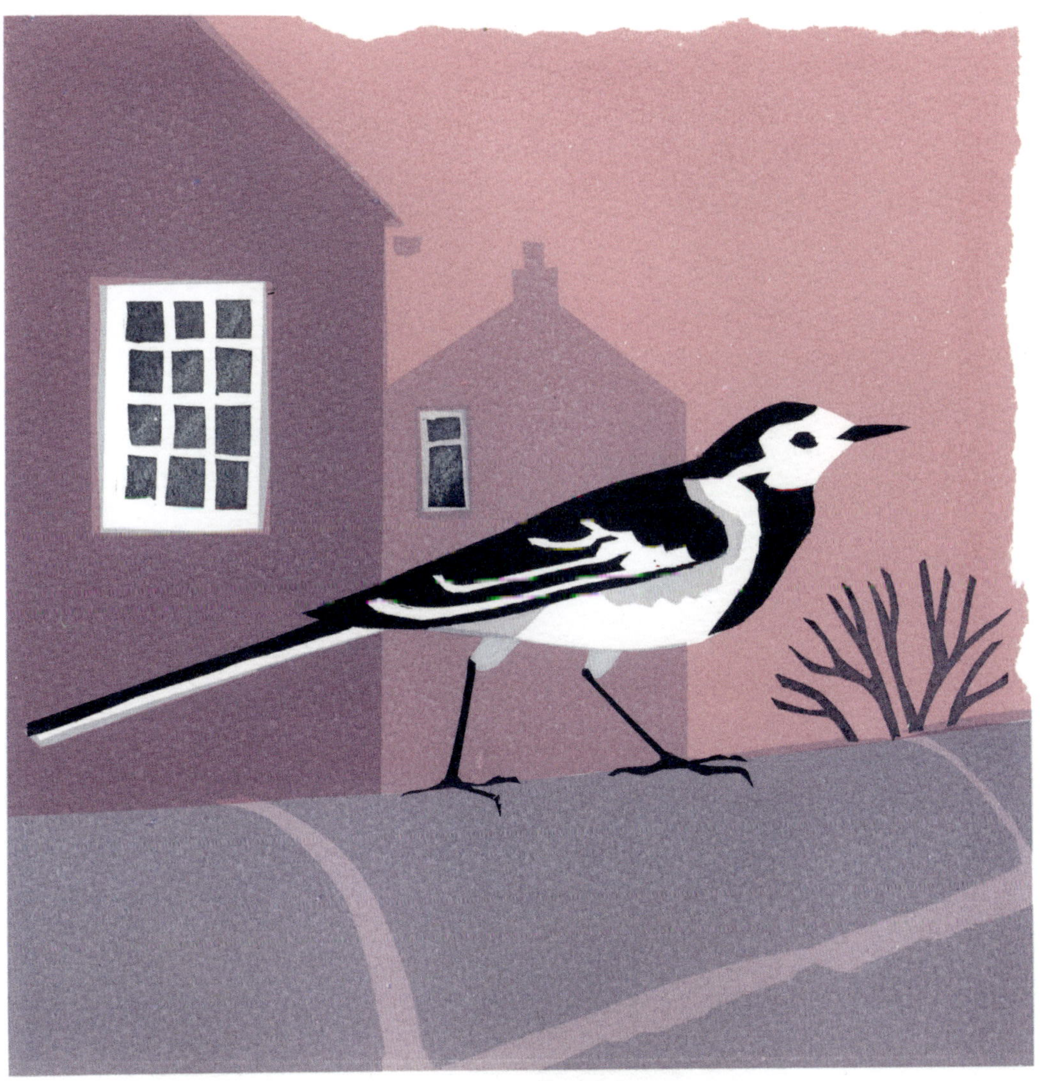

The **pied wagtail** (*Motacilla alba yarrellii*) announces its presence with its *chizzik* call. It appears in the most surprising places, darting in and out regardless of the heedless steps of rush-hour commuters or the flying feet of children playing on a beach. One appeared on Wimbledon's Centre Court during a Djokovic match. In winter, they roost by the thousands for warmth in a range of venues. An analysis of Poole Harbour, Dorset, revealed them congregating in an oil-gathering station, trees around a supermarket car park and yachts in a marina. Heathrow terminals are another favourite roost. Two hundred roosted in Bridgend's Christmas tree (2020).

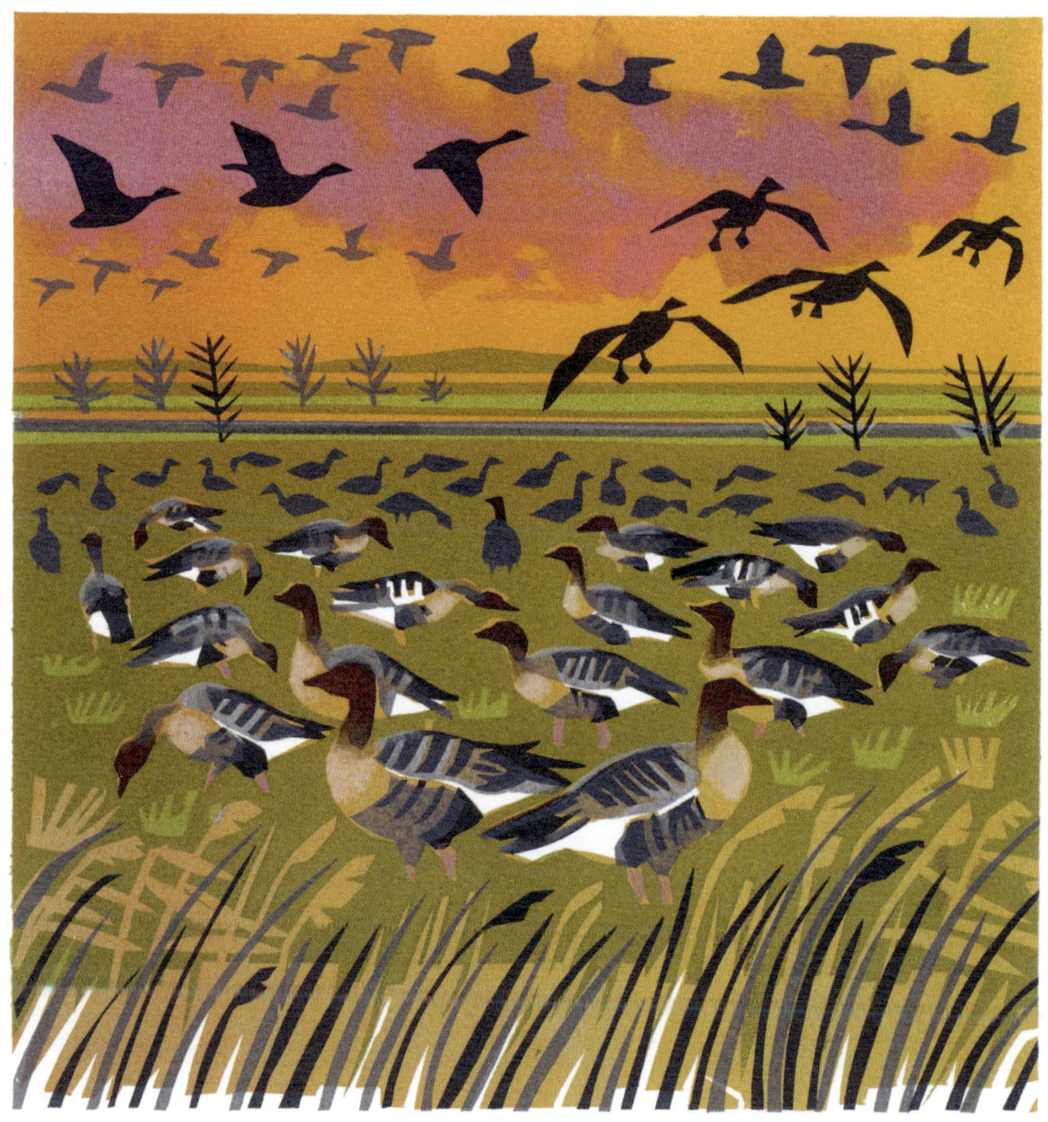

Skeins of the **pink-footed goose** (*Anser brachyrhynchus*) in full cry are well-named 'the hounds of heaven'. Discarded sugar-beet tops have supplemented its diet to dramatic effect: 1.5m winter here from Iceland, on arrival most numerously in the Montrose Basin, later in the Wash and north Norfolk.

And far above the Angus straths I saw the wild geese flee,
A lang, lang skein o' beatin' wings wi' their heids towards the sea,
And aye their cryin' voices trailed ahint them in the air –
O wind, hae mercy, haud yer wisht, for I daurna listen mair.

Violet Jacob (1863–1946), from 'The Wild Geese'

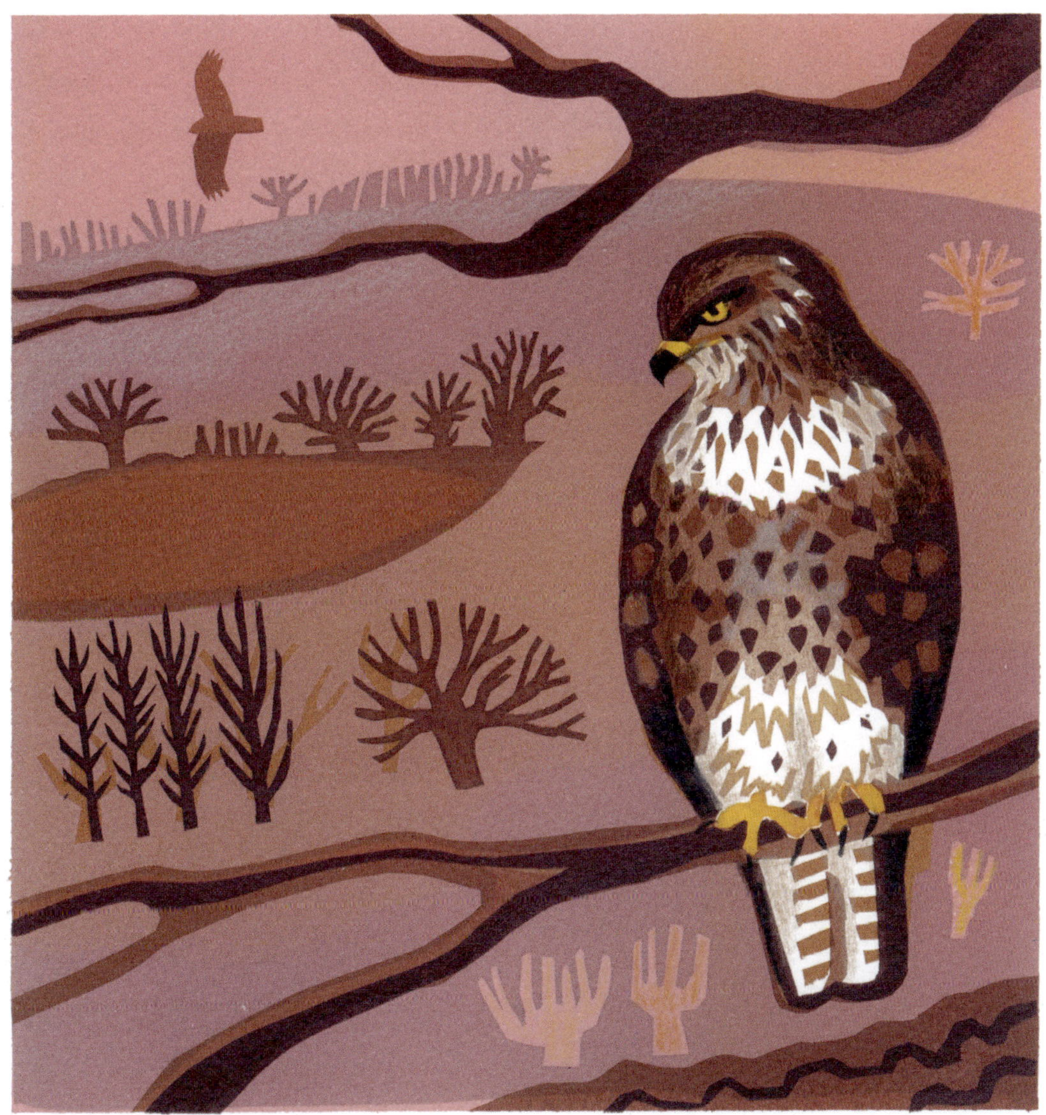

Once shot and trapped as vermin almost to extinction, the **buzzard** (*Buteo buteo*) is now the UK's commonest bird of prey. A pronounced move eastward from Wales means buzzards have even reappeared on London's Hampstead Heath. While their soaring spring aerobatics are mesmerising, when perched, their stillness and brown camouflage render them invisible despite their size. If they take wing, a roadside pole or branch alarmingly springs to life.

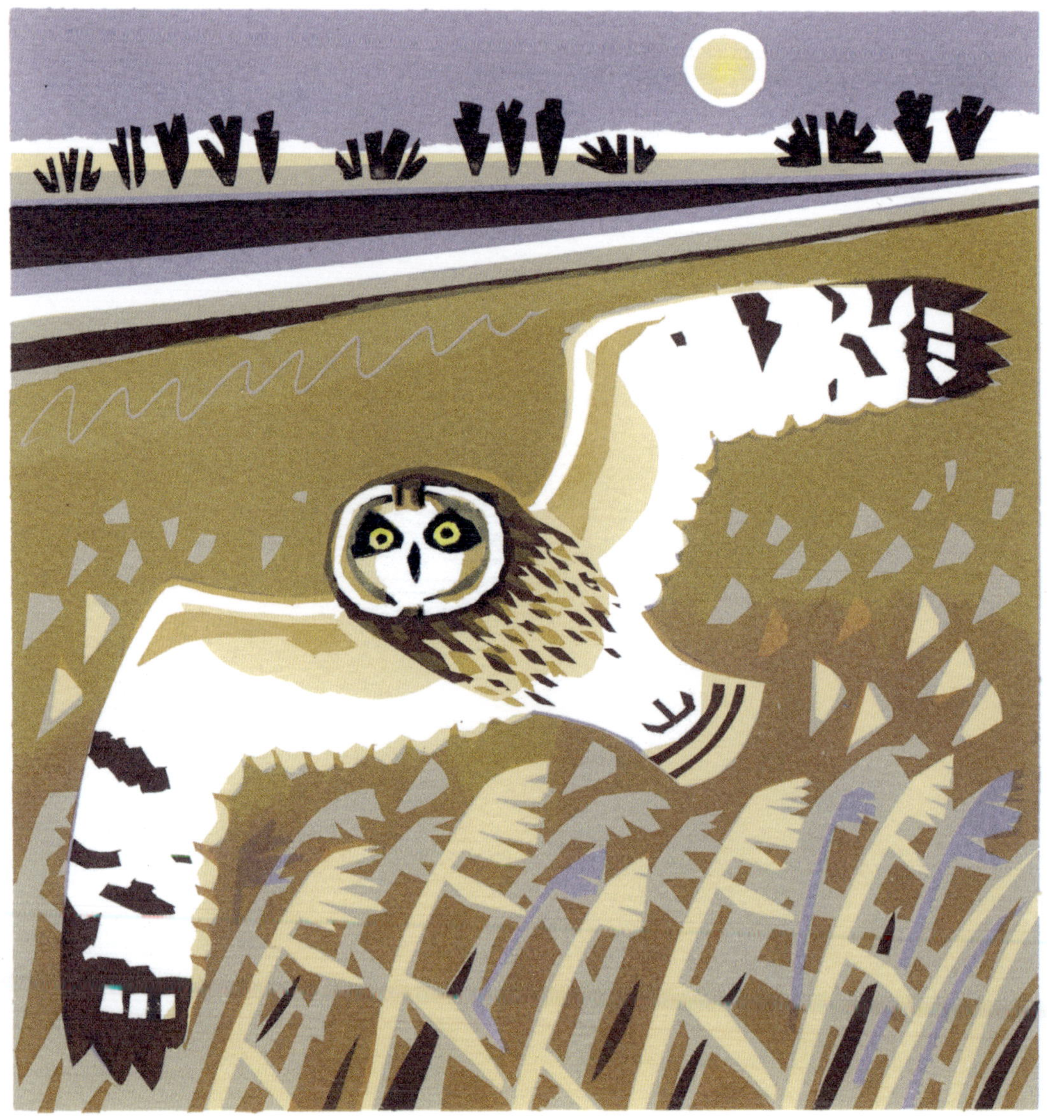

The tufts that give the **short-eared owl** (*Asio flammeus*) its name are for display, not hearing; the same applies to the nocturnal and wood-dwelling long-eared owl (*Asio otus*). The short-eared is notably diurnal. Winter marshes offer daylight sightings, not least from the east-coast train line to Scotland as it passes the Great Fen south of Peterborough. A significant migration from Europe – the number unestimated – joins the UK's 2,200 pairs, some of which may emigrate, with food abundance dictating unforeseeable wanderings. The highest-flying owl I have seen was a short-eared at Welney Nature Reserve in Northamptonshire. One minute perched close by, the next a diminishing spot with two soaring buzzards.

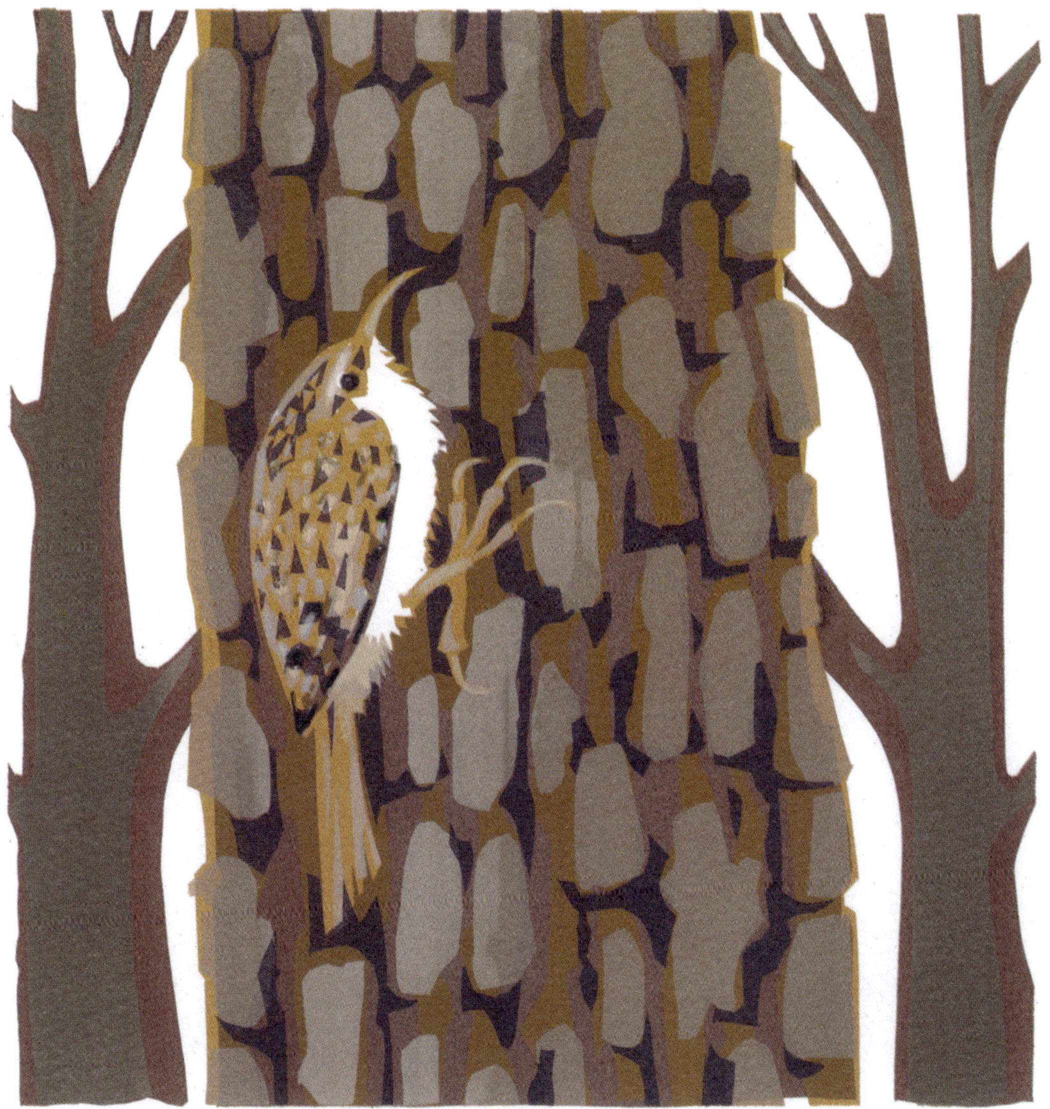

The **treecreeper** (*Certhia familiaris*), most mouse-like of birds, ascends tree trunks for food; the other trunk-creeping bird, the nuthatch, ascends and descends. Nuthatches visit bird tables, while the insect-dependent treecreeper sticks to trees. Actor John R. H. McEwen writes: 'I like the treecreeper for being such a perfect little package of form and function.' He reports two nests at head height in an old wooden Berwickshire barn. They customarily nest under bark, where they are prey to great spotted woodpeckers. The soft and chunkily-barked giant redwood (*Sequoiadendron giganteum*) is their favourite all-purpose tree. The easily worked bark provides for nests, and a peppering of insulated holes can reveal a roosting group.

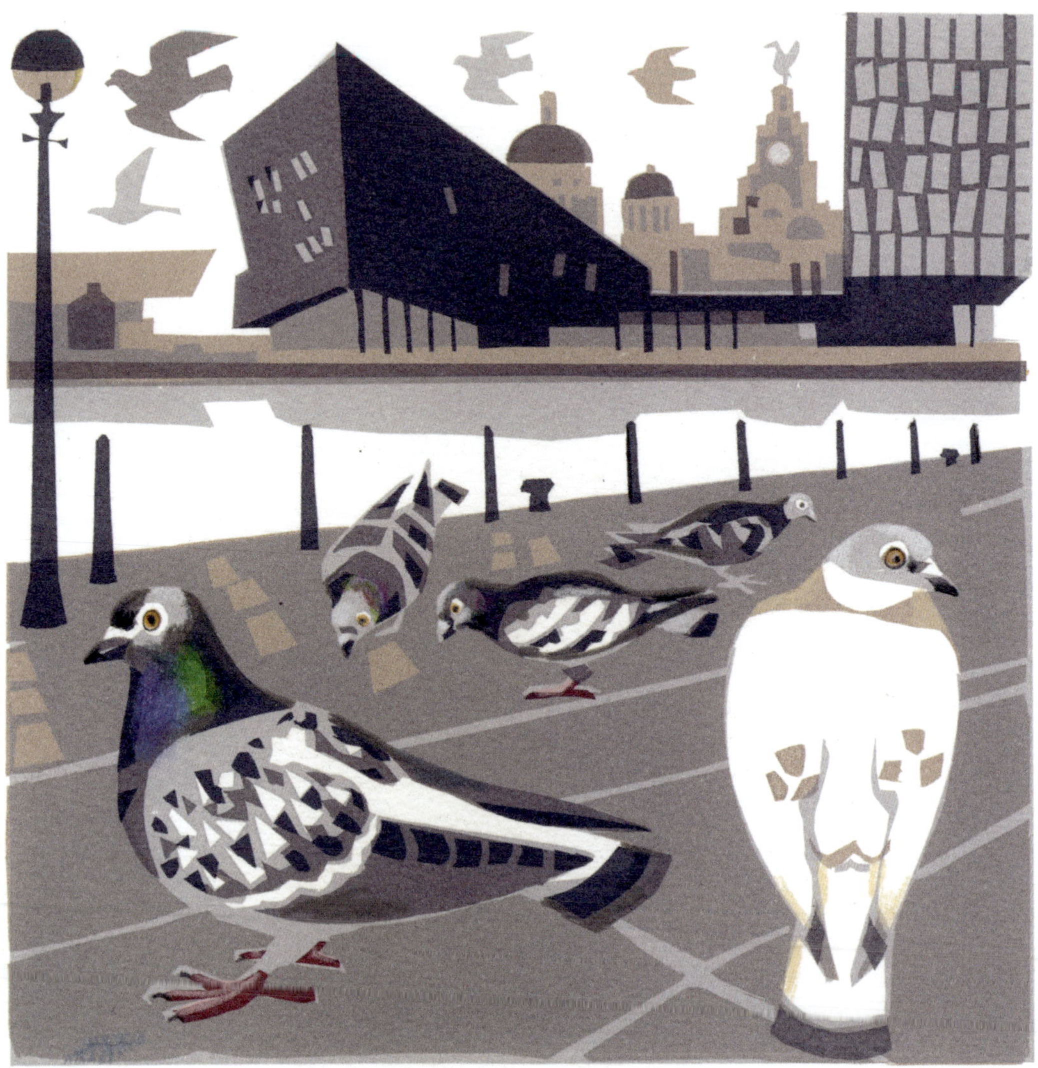

The pigeon is the longest-domesticated bird; its numerous breeds derived over 10,000 years from the cave-dwelling rock dove. The varied genes of domestic pigeons meet in the **feral pigeon** (*Columba livia domestica*), found globally. Adrian Richmond tends the flock on his *Big Issue* pitch outside London's Westminster Cathedral. He signed his Christmas card from 'Adrian, Tanglefoot, Peanut, Tricia, Dixie, Maria & Mrs Whitebird, Sydney, Stanley, Cpt Beattie & All the gang'. In 2023, one of the flock was loose in the Cathedral for six days. On release, too weak to fly, it was protected for the night by Tracy Wood, who shared her pavement bedding. 'Captain Beattie', who 'never gets his feathers wet', is the oldest (12) of Adrian's favourites. '*They* aren't the "rats". *We* are,' he says. Marilyn Monroe agreed. She paid New York urchins to release the pigeons they caught.

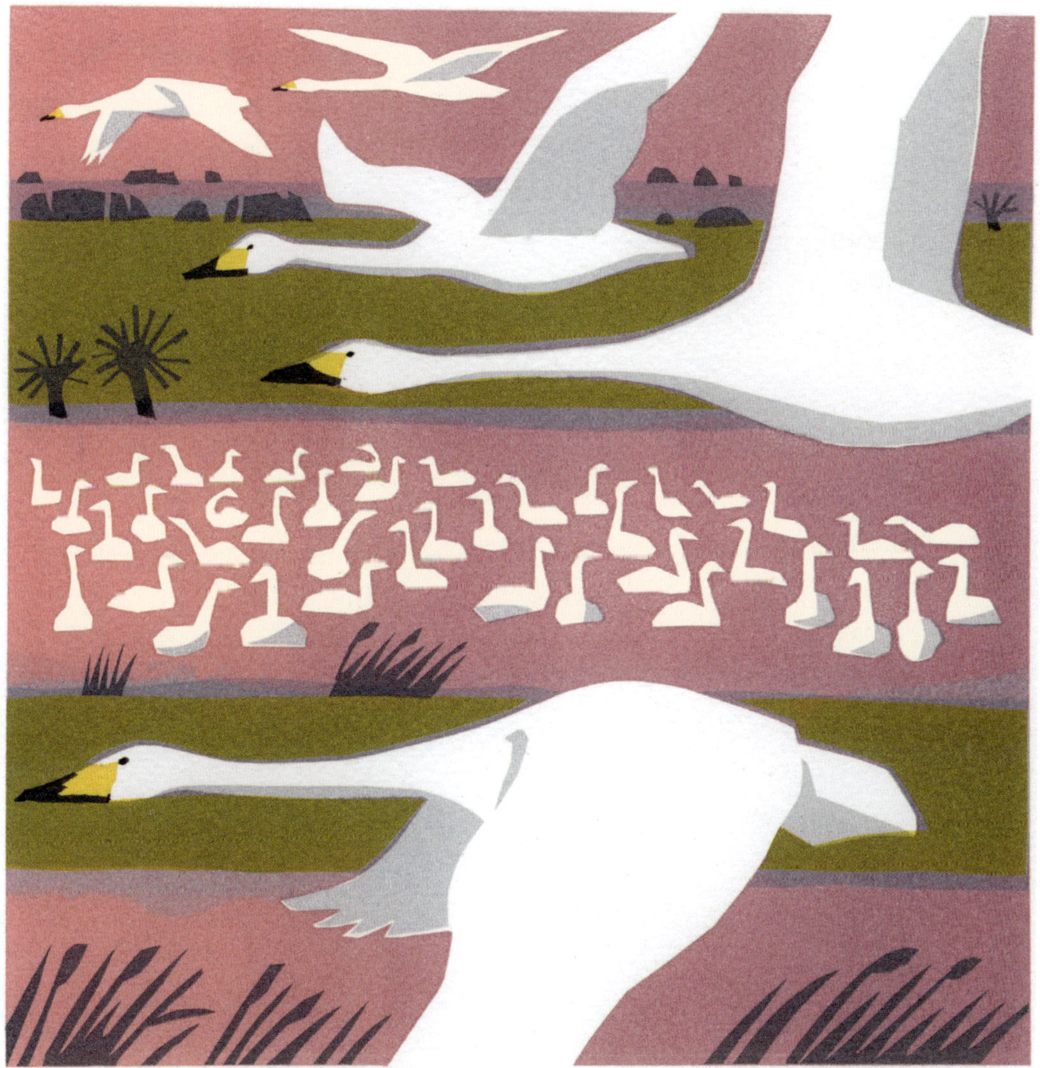

In winter, two species of wild swan migrate to the UK, the **whooper swan** (*Cygnus cygnus*) and smaller Bewick's. The Anglo-Saxon *whopan*, 'cry out', gives the whooper its name. It can complete the 966km (600 miles) from Iceland in under a day and withstand temperatures of -50°C. In December 1967, radar in Northern Ireland revealed a skein flying at 8,230m (5 miles), making it the world's fourth highest-flying bird. Like ducks and geese, they feed by night or day according to moon and tide. Swimming, they do not raise their wings like sails as the mute does, and they cause less of a commotion than the mute when taking off. In flight, their wing beats make a breathless sound, the mute's a throb. Feeding on seagrass, they are meticulous, whereas mutes dredge up and swallow beakfuls.

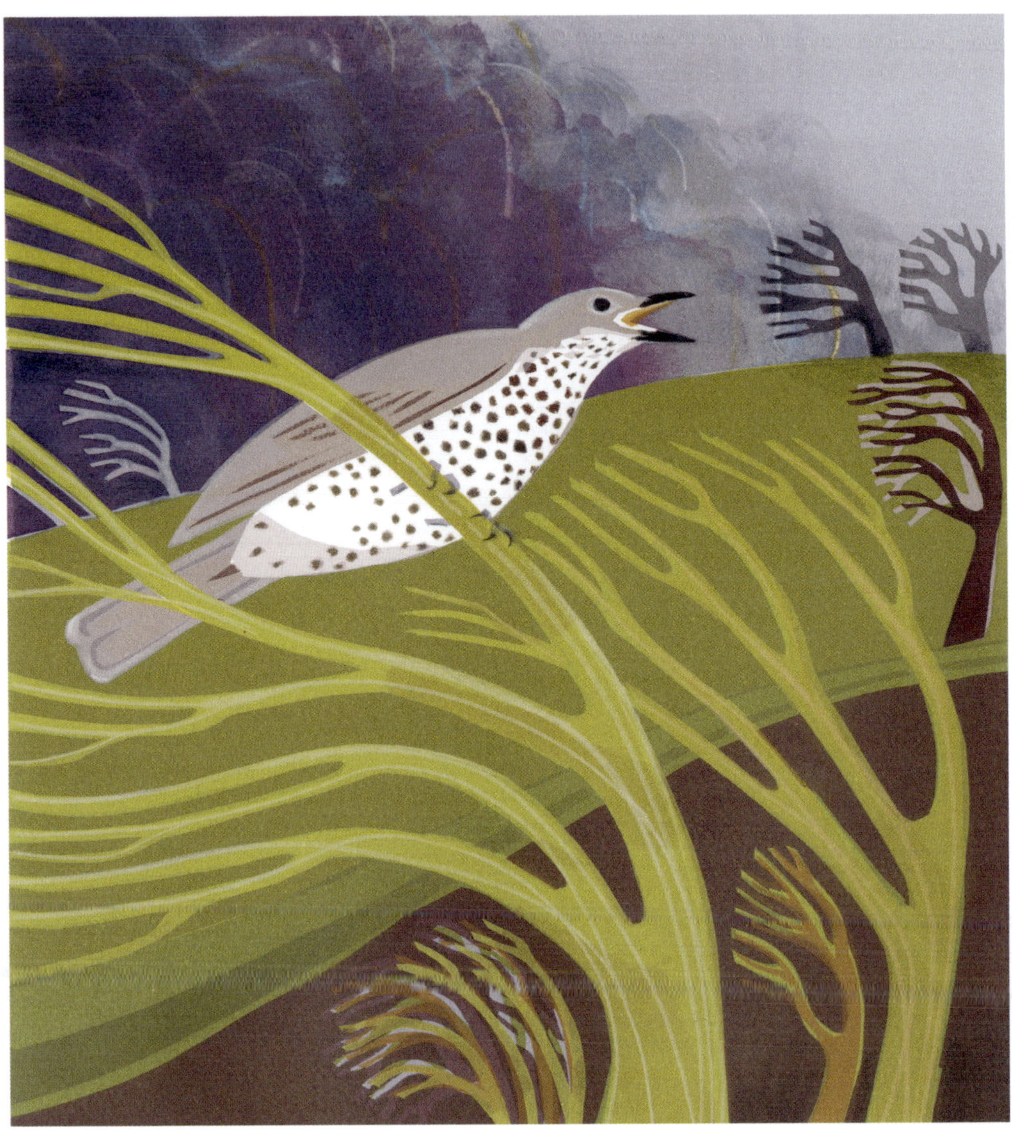

The **mistle thrush** (*Turdus viscivorus*) owes its English name, derived from Aristotle, to Sir Thomas Browne. 'Mistle' is short for mistletoe, the evergreen parasitic plant Druids revered for its healing properties and that we hang to kiss under at Christmas. *Viscivorus* means 'devourer of mistletoe'. In autumn the mistle thrush is most noticeably seen in flocks, flying or on pasture. Of its many colloquial names, 'storm cock' is earned for its singing whatever the weather, even snowstorms. As Thomas Hardy wrote in 'The Darkling Thrush', it is as if:

Some blessed Hope, whereof he knew
And I was unaware.

A pair nested in a Leeds traffic light (2020), the young hatching during Storm Dennis.

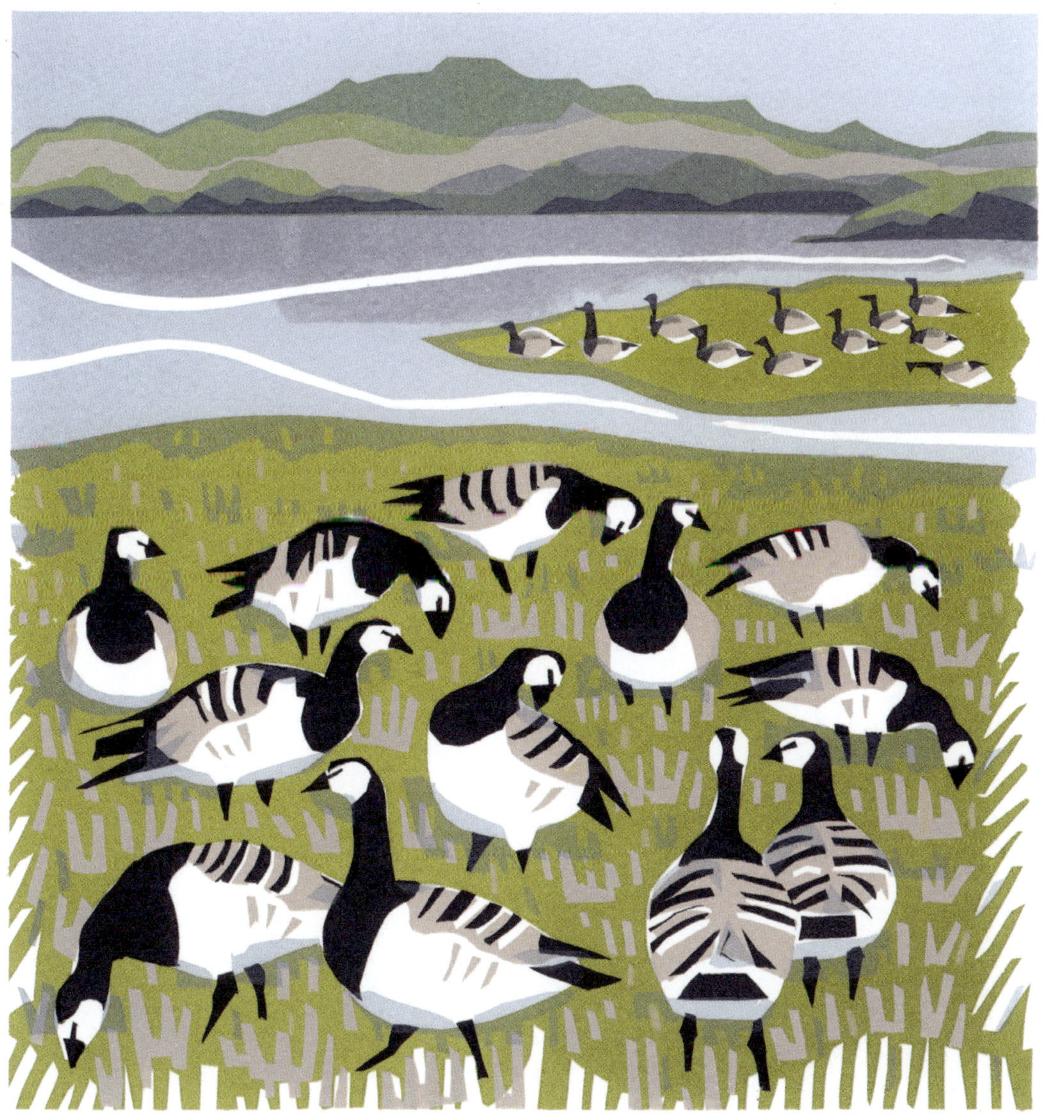

The BBC's *Life Story* film of a **barnacle goose** (*Branta leucopsis*) gosling, jumping from its nest ledge 122m (400ft) up a Greenland cliff has accumulated 48m hits on YouTube. Its paddle-steered glide ended after the last of several jolting contacts with the rock face, leaving its body a limp rag at the foot. And yet it survived. Tradition said the barnacle goose emerged from the shellfish, lent authority by priest and historian Giraldus Cambrensis, 'Gerald of Wales' (1146–1223): 'I have frequently seen ... more than a thousand of these small birds, hanging down on the sea-shore from one piece of timber.' The number of UK-wintering barnacle geese is 105,000. Deaths from avian flu in 2021–2023 reached 5,000 on Islay and 13,200 in the Solway Firth.

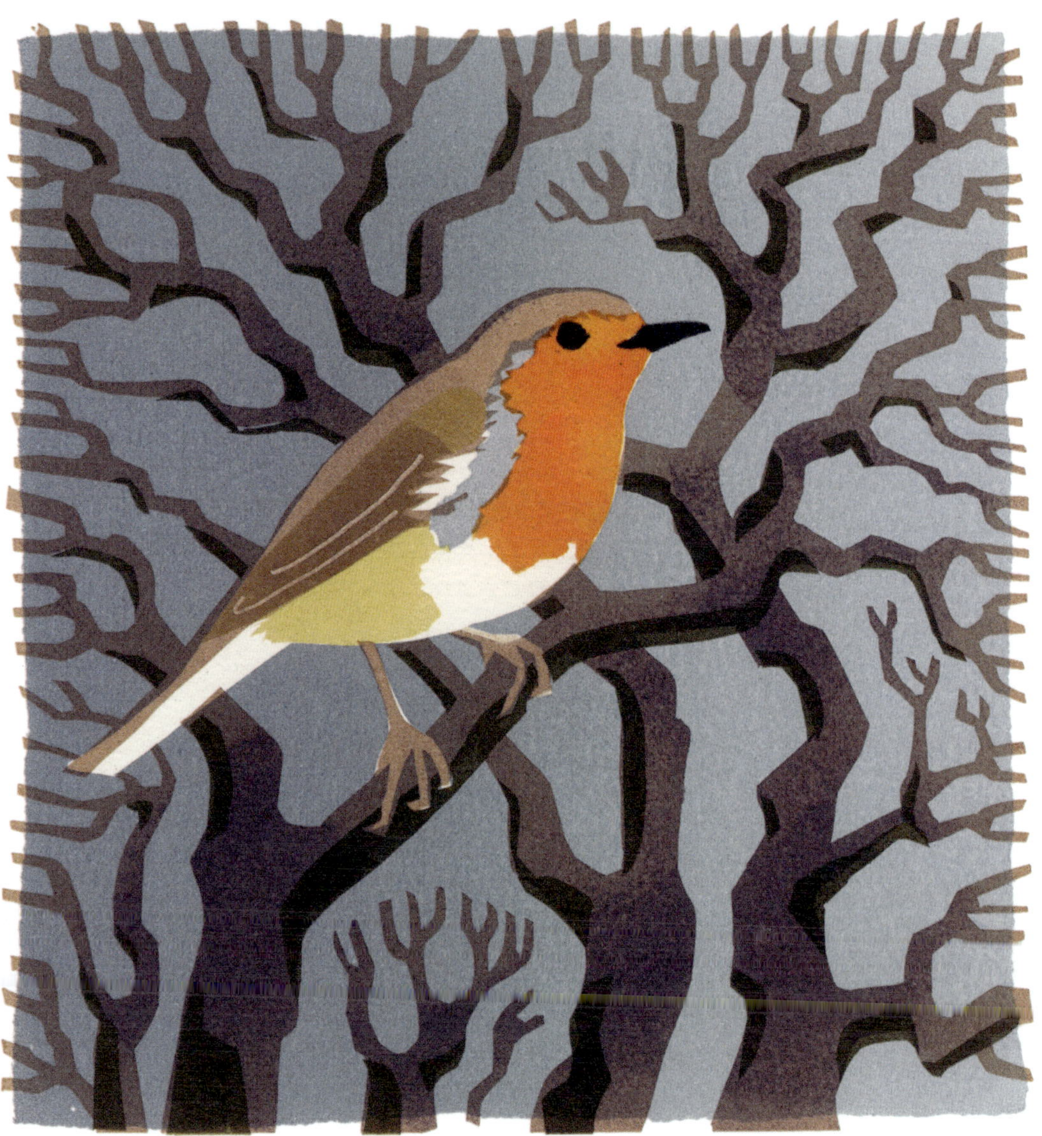

The **robin** (*Erithacus rubecula*), formerly classified a thrush (Turdidae), has been reclassified a flycatcher (Muscicapidae). Despite its ability to fight to the death for territorial control, this is usually achieved by out-singing a rival. The song sounds ecstatic in spring, wistful in autumn. Streetlights encourage nocturnal and winter singing in town. It is a ready presence when digging promises pickings. Lord Grey wrote and demonstrated that 'any male robin can be tamed'. It has long had sacred associations:

The robin redbreast and the wren
Are God Almighty's cock and hen

Anon, Traditional

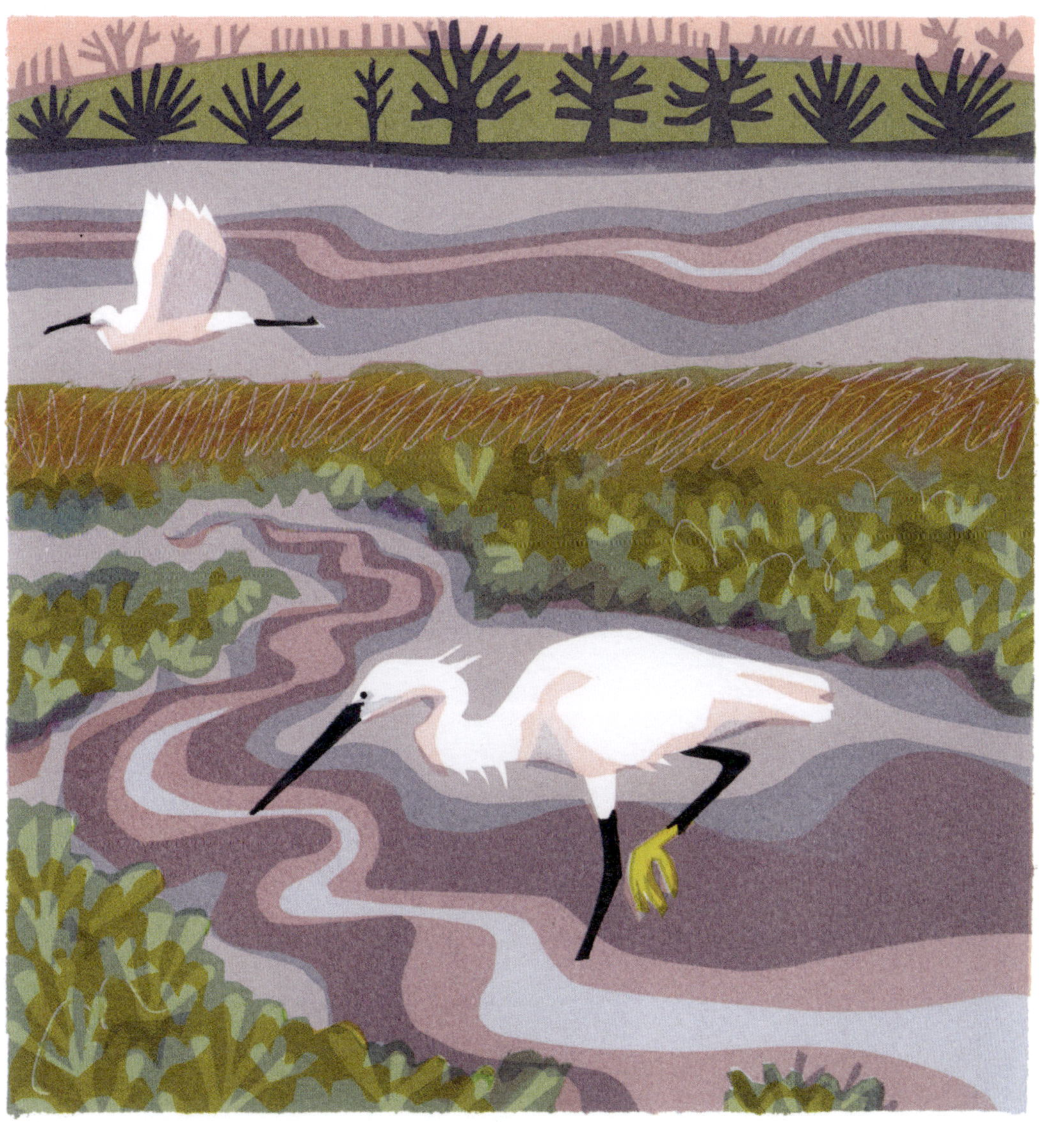

The brightest light by rivers, lakes, marshes and creeks is the white of the **little egret** (*Egretta garzetta*). A banquet for the 1465 enthronement of an archbishop of York included 1,000 little egrets. By the early nineteenth century, Thomas Bewick despaired that they were almost extinct. A fashion for feather-ware accounted for their slaughter abroad. In 1886, one London dealer marketed a million skins. The first recorded modern breeding in the UK was at Dorset's Brownsea Island Nature Reserve in 1996. Now, there are 1,100 resident UK pairs of this tree-nesting heron and a winter population enlarged by 10,500 migrants. The smaller cattle egret (15 UK breeding pairs) and larger great white egret (12 UK breeding pairs) both increase in population in winter by about 60 migrants.

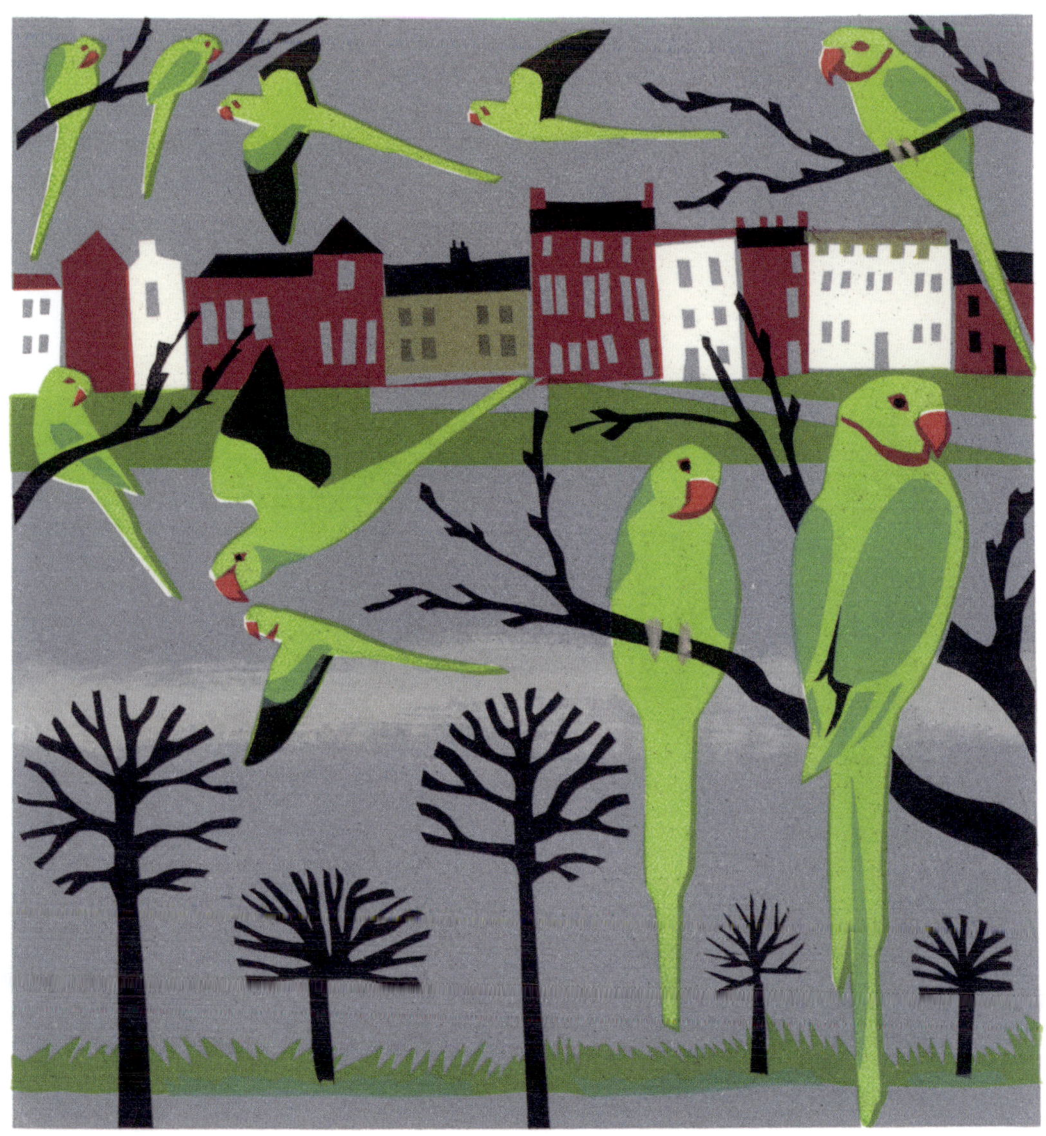

The apparent novelty of the **rose-ringed parakeet** (*Psittacula krameri*), listed as a British bird in 1983, is offset by its long association as a cage bird. John Skelton (1463–1529), in his poem 'Speke Parrot', described its rose-ringed narrator:

> *My fedders fresshe as ys the emrawde grene,*
> *Abowte my necke a cyrculet lyke the ryche rubye...*

Its expansion as a UK resident dates from a first wild nesting at Croydon in 1971; it is now widespread in London and the south-east. Officially, the UK has 12,000 pairs, which seems conservative considering there are populations as far afield as Dundee and Ireland. Regarded a pest in India, it may threaten the English wine industry as it is said to relish vineyards.

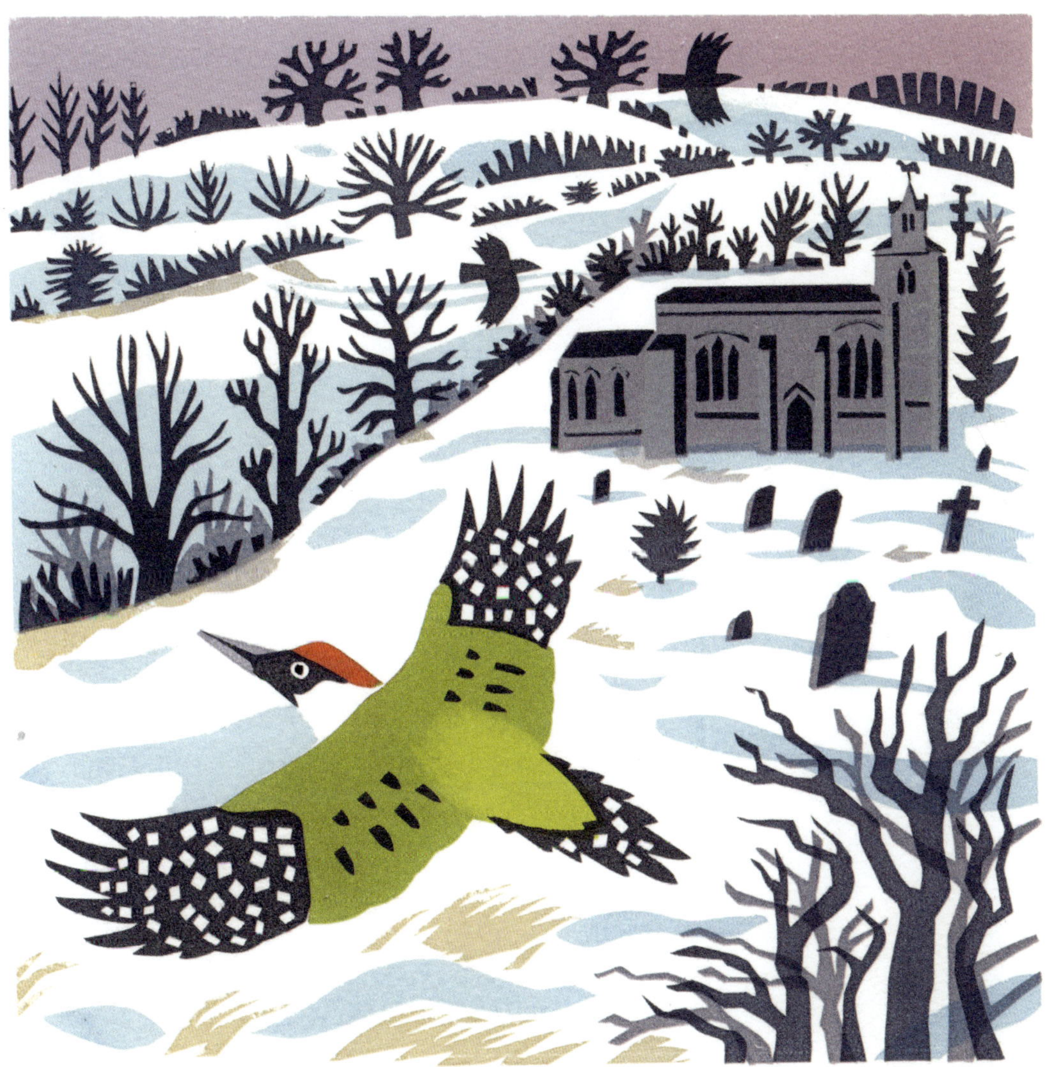

The **green woodpecker** (*Picus viridis*) is readily heard – a piping *kew-kew-kew* earning it its nickname 'yaffle' – and seen on garden lawns and urban parks, ants being its principal food. Ant nests are underground. It probes with a sticky 10cm-long (3.9in) tongue that coils back inside its skull. Its predominant green makes it surprisingly invisible when feeding. Frozen ground is its enemy.

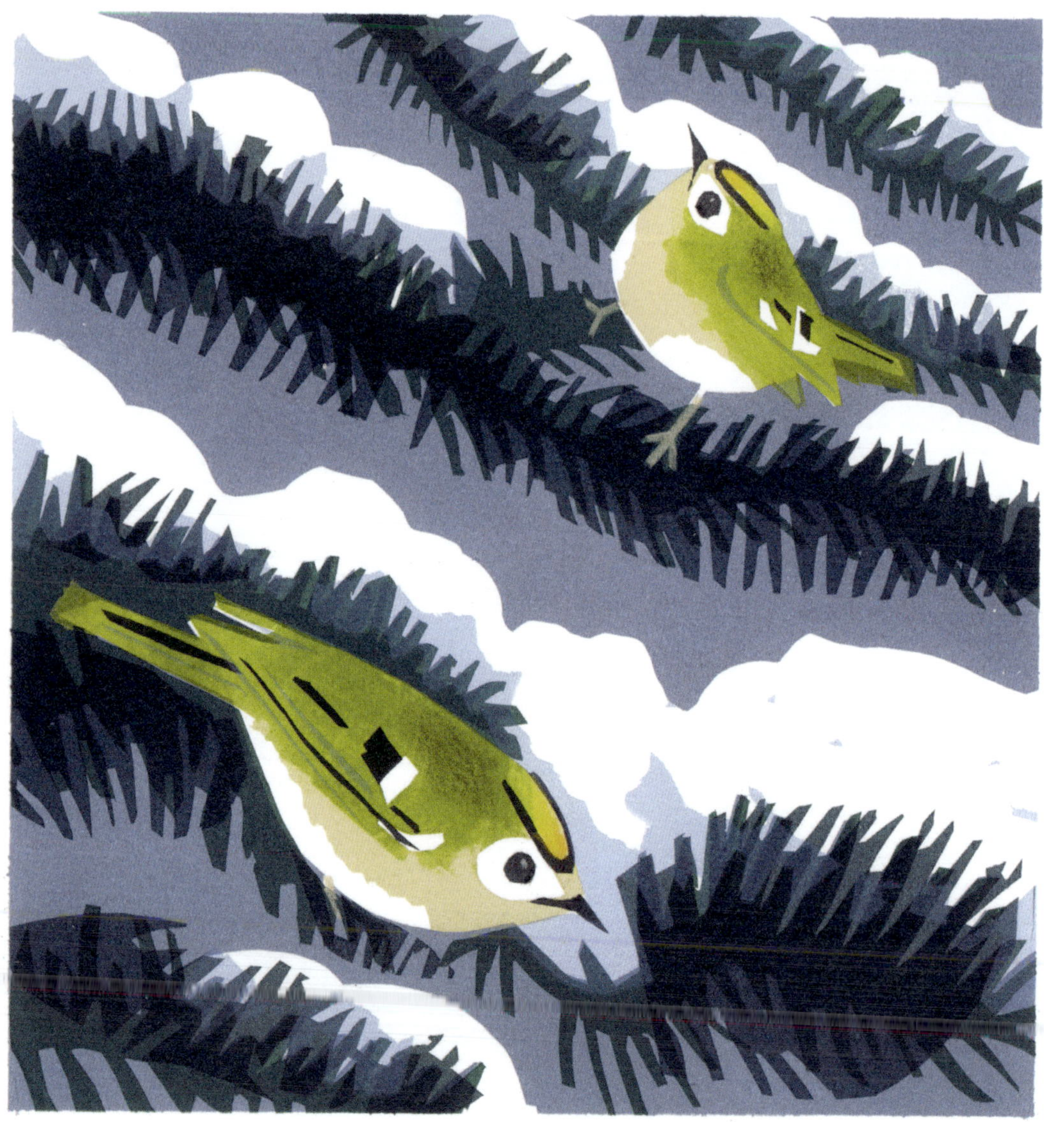

The **goldcrest** (*Regulus regulus*) derives its name 'kinglet' (*regulus*) from Aristotle's fable of how this smallest of birds robbed the soaring eagle of the title 'King of Birds' by hiding in the eagle's feathers and fluttering free at the zenith to be crowned by the Sun. Henry Douglas-Home wrote of its 'nest of moss and feathers, the size of half a tennis ball, slung with cobwebs to hang like a hammock' at the tip of a lower branch of an old yew tree. However strong the wind, the nest remained unbroken, 'while the heavier constructions of other birds are frequently dumped by a gale' (*The Birdman*). The only other European bird to hang its nest is the golden oriole, which last bred in the UK in 2009.

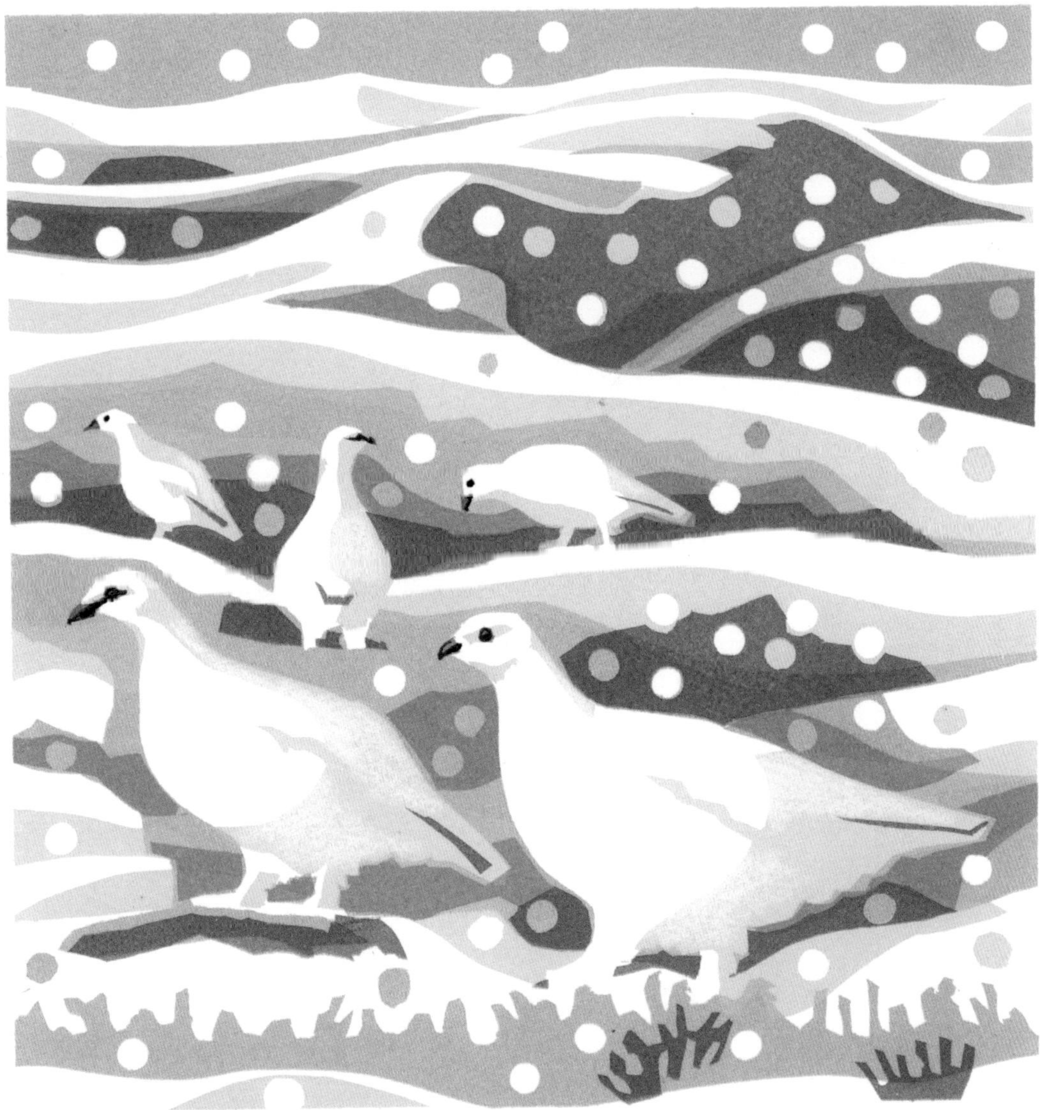

The name **ptarmigan** (*Lagopus muta*) comes from Gaelic *tarmachan*, the 'pt' added by physician and antiquary Sir Robert Sibbald (1641–1722) to make it sound more scholarly and Greek (e.g. pterodactyl). It is one of three British species (along with the stoat and mountain hare) to turn white in winter and has the highest habitat, its moss and lichen dependence rarely taking it under 750m (2,400ft). Unused to humans, it can be seen by Aviemore skiers within sight of the Ptarmigan Restaurant in the Cairngorms. It has feathered soles and nostrils and burrows in snow for food and warmth.

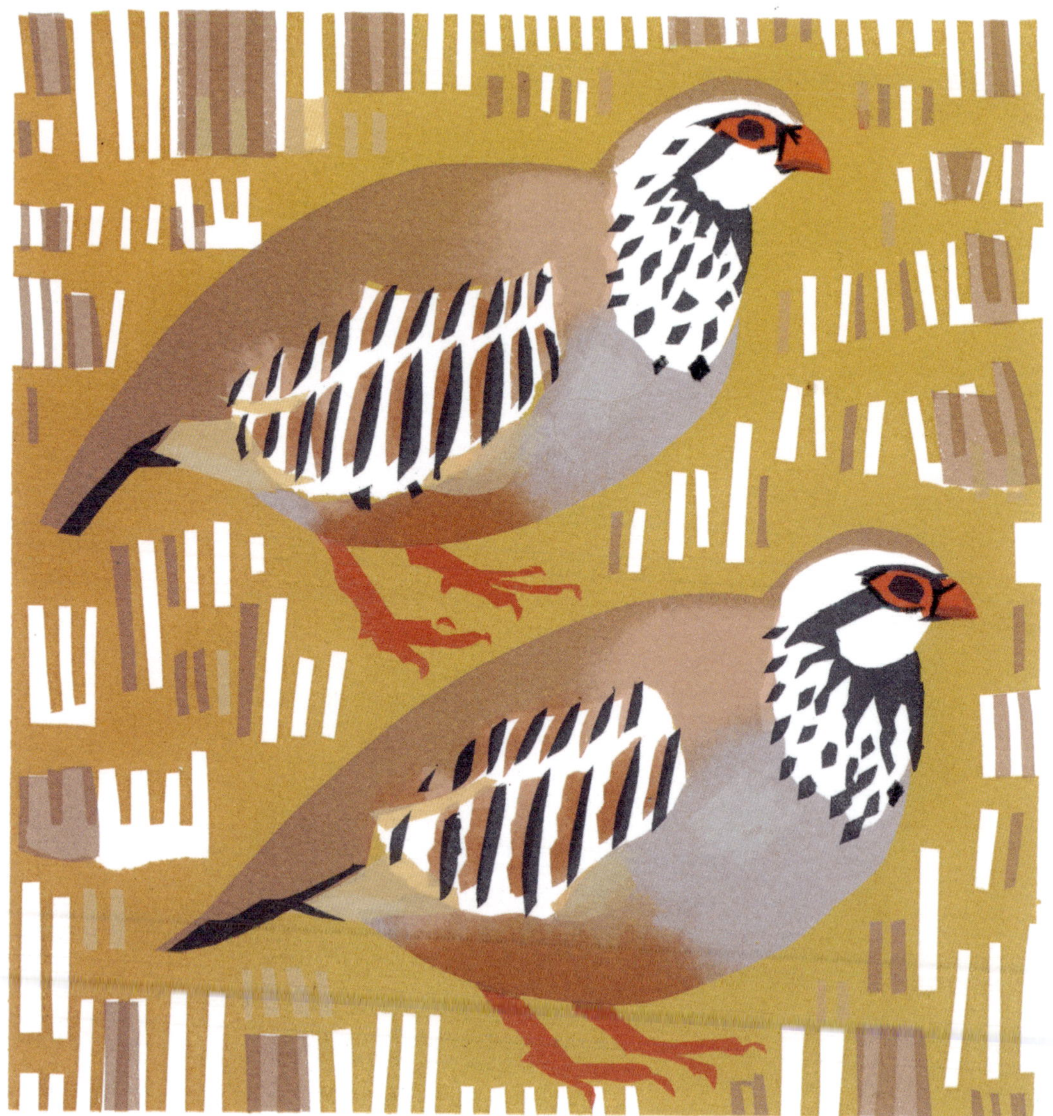

On the first day of Christmas, my true love sent to me
A partridge in a pear tree

From 'The Twelve Days of Christmas'

The partridge of the carol's refrain is a red-leg (*Alectoris*). In the British Isles, the species is the **red-legged partridge** (*Alectoris rufa*) or 'French' partridge, introduced from France as a game bird around 1770. The red-legged is larger, bolder plumaged, less squat than its 'grey' 'English' cousin. More than 4m red-leggeds are reared annually for UK commercial shoots. In the carol, *pear tree* is a nonsensical misapplication of French *perdrix*, 'partridge', pronounced 'peardree'.

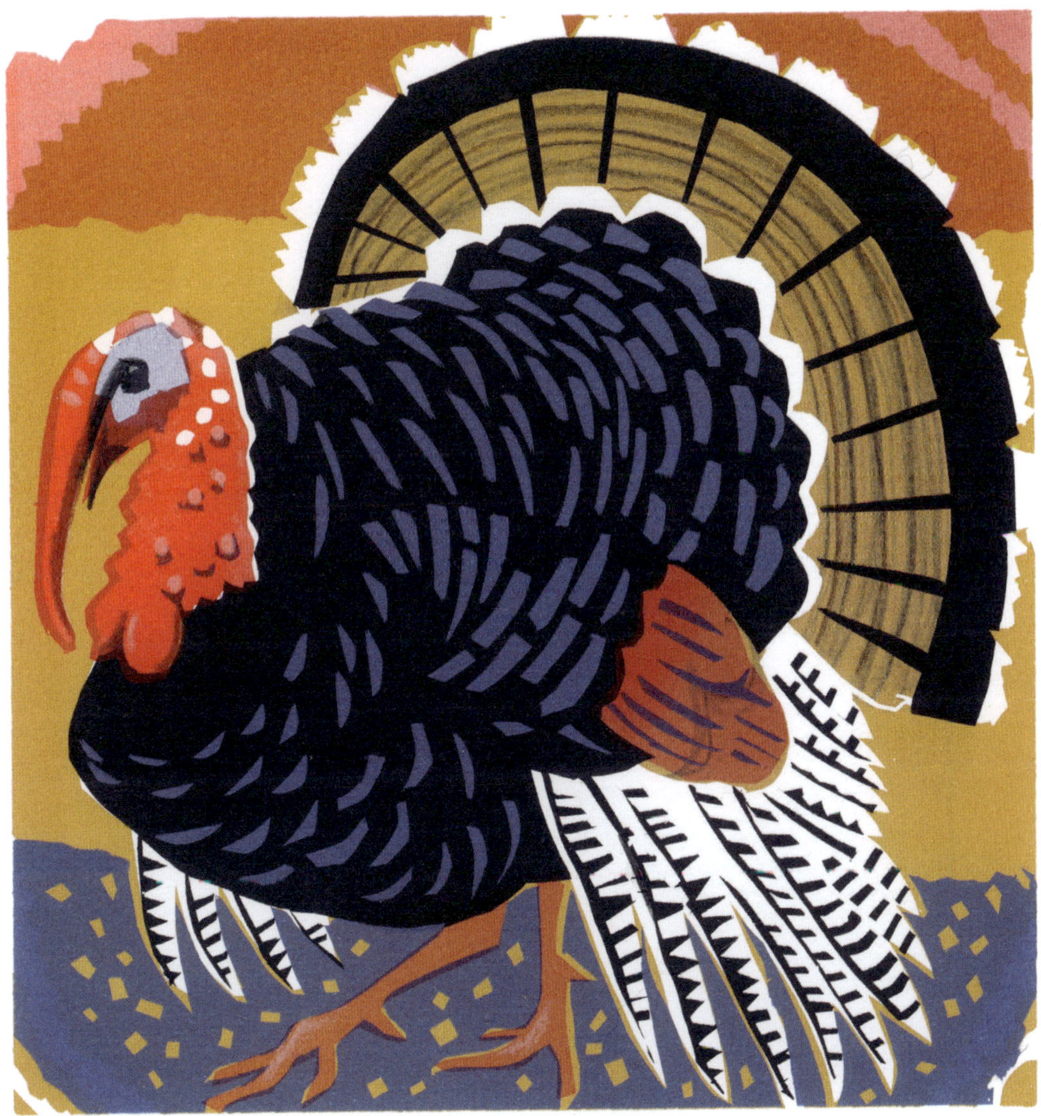

The wild American **turkey** (*Meleagris gallopavo*) is legally hunted throughout the US. King George II (1689–1760) was first in failing to establish it as a British game bird. Its English name dates from its sixteenth-century import as a table bird via the Turkish Levant, where European traders had settled in the coastal cities of the Ottoman Empire. The Christmas turkey replaced the goose in the eighteenth century. The present century boasts 40 farmed breeds. In the UK, we eat 10m turkeys at Christmas, except in 2022 when, being farmed free range, the number was diminished by avian flu, most damagingly in Norfolk, centre of the industry, with some farms closing.

ACKNOWLEDGEMENTS

To James Pembroke, Harry Mount (and for his foreword), Penny Phillips, Michael Hardaker and all at *The Oldie*. To Richard Ingrams, founding editor, for commissioning 'Bird of the Month'. For permission to publish: Swarden Burn ('Pinkfeet'), Katharine Reid ('Bird Watching'), Peregrine Douglas-Home (*The Birdman*, Collins), Richard Smyth (*A Sweet Wild Note*, 2017, Elliott & Thompson). To Henry and always-lamented Tessa Keswick (Neil Annan, head keeper) and Matt and Anya Ridley for visits to the unmatched havens for nesting waders and black grouse on their Angus and Durham grouse moors. For access to other vital private conservation projects - Dorset: Ned Cranborne, Cranborne (Beau Witney, keeper). Essex: Jonny and Miff Minter, Purdey Award-winning Rivers Hall (Robert Graves, keeper). Hampshire: Christopher Wills, Litchfield. Kent: Gareth Fulton, for a tour of his privately owned but public Elmley Nature Reserve; Vyvyan and Alexandra Harmsworth, David Lodge and Birty, Stodmarsh Nature Reserve. Norfolk: Ollie Birkbeck, Little Massingham; Andrew Crean, Deepdale Marsh (Mark Andrew, marsh warden); Jake Warman for loaning his cottage; his father Alister (RIP), bird-loving angler and art supremo. Northumberland: Alex and Natasha McEwen, jollity/bird reports. Shropshire: Gareth Thomas. South Uist: Mary Miers, corncrakes, golden eagles. Wiltshire: Simon and Jenny Courtauld, great bustards. Rose Yorke for alerting me to Yvette Brown. Paul Langley and daughter Chloe for calendar support (Cramer's Butchers, Islington). David Nawaraukas, Adrian Richmond and Tracy Wood. Rose Ryan for inspired idiosyncrasy (e.g. Henry Longhurst, *My Life and Soft Times*). Books: Mark Cocker, *Birds Britannica*; Stephen Moss, Brent Westwood, *Tweet of the Day*; Bryan Holden, 'BB' anthologies; Francesca Greenoak, *British Birds* (bird names); Ian Coghill, *Moorland Matters*. For bird news: Duncan, David, Jo, Oscar and Helena McEwen; Helen Bellany, Steve Chong (computer wizard), Cecilia Cole, Penelope Enthoven, Christabel Holland, Sandra Lawrence, Badger Walker. To the staff of the London Library, and above all to Bloomsbury for commissioning the book, especially Julie Bailey, Publisher, Kate Dickinson, Assistant Editor, and also Austin Taylor, designer. For 'Birds': Gordon Monk for lots of stuff, Steve Brayshaw for ornithological consultation. For 'Words': Katharine Reid without whom there would not have been any. For any oversight, we trust you know the authors well enough not to mind.

INDEX